D1445299

ECONO

IN A WARMING WORLD

Since the publication of the Stern Review, economists have started to ask more normative questions about climate change. Should we act now or tomorrow? What is the best theoretical carbon price to reach long-term abatement targets? How do we discount the long-term costs and benefits of climate change?

This provocative book argues that these are the wrong sorts of question to ask because they don't take into account the policies that have already been implemented. Instead, it urges us to concentrate on existing policies and tools by showing how the development of carbon markets could dramatically reduce world greenhouse-gas emissions, triggering policies to build a new low-carbon energy system while restructuring the way agriculture interacts with forests. This provides an innovative new perspective on how a post-Kyoto international climate regime could emerge from agreements between the main greenhouse-gas emitters capping their emissions and building an international carbon market.

CHRISTIAN DE PERTHUIS is a professor of economics at the University Paris-Dauphine, where he is also the Director of the Masters programme in Energy, Finance and Carbon, and a member of the Council on Economics and Sustainable Development, a body advising the French Minister for the Environment. His most recent book is *Pricing Carbon* (Cambridge University Press, 2010), co-edited with A. Denny Ellerman and Frank Convery.

Economic Choices
in a Warming World

Christian de Perthuis

CAMBRIDGE
UNIVERSITY PRESS

CAMBRIDGE UNIVERSITY PRESS
Cambridge, New York, Melbourne, Madrid, Cape Town,
Singapore, São Paulo, Delhi, Tokyo, Mexico City

Cambridge University Press
The Edinburgh Building, Cambridge CB2 8RU, UK

Published in the United States of America by Cambridge University Press, New York

www.cambridge.org
Information on this title: www.cambridge.org/9780521175685

Authorized translation from the French language edition,
entitled *Et pour quelques degrés de plus*, by Christian de Perthuis,
published by Pearson Education France,
© Christian de Perthuis 2009.

First published in French 2009
English edition first published 2011

Printed in the United Kingdom at the University Press, Cambridge

A catalogue record for this publication is available from the British Library

Library of Congress Cataloguing in Publication data
Perthuis, Christian de.
[Et pour quelques degrés de plus. English]
Economic choices in a warming world / Christian de Perthuis.
p.   cm.
First published in French 2009.
ISBN 978-0-521-17568-5 (pbk.)
1. Global warming–Economic aspects.   2. Climatic changes–Economic
aspects.   3. Environmental policy–Economic aspects.   4. Greenhouse gas
mitigation.   I. Title.
QC981.8.G56P475   2011
363.738′74–dc22
2010051872

ISBN 978-1-107-00256-2 Hardback
ISBN 978-0-521-17568-5 Paperback

For Manon and Cloé,
In memory of Claire

The English version of this book is the translation of the second edition of
Et pour quelques degrés de plus, *translated by Michael Westlake,*
with the assistance of Natalie Frank and the financial backing of the
Finance et Développement Durable (FDD) Chair, sponsored by
Electricité de France (EDF), Crédit Agricole-CIB and the
Caisse des Dépôts.

Contents

Acknowledgements

Economic Choices in a Warming World is an English translation of the second edition of *Et pour quelques degrés de plus* that has been completely reworked. The structure of the book has been reorganized and a large part of it rewritten. I was encouraged to make these changes by the reactions of readers to the first edition, and it is to them that my initial thanks are due: their feedback was invaluable.

Through the support of the Caisse des Dépôts – whose directors, beginning with the general manager Augustin de Romanet, I warmly thank – I joined the Finance et Développement Durable (FDD) Chair, attached to Paris-Dauphine University and the École Polytechnique, and was able to launch the new Climate Economics Chair, a joint initiative by CDC Climat and Paris-Dauphine University under the aegis of the Europlace Finance Institute Foundation. I would also like to thank the directors of the FDD Chair for their welcome, Pierre-Louis Lions and Jean-Michel Lasry; the directors of the Centre Géopolitique de l'Énergie et des Matières Premières (CGEMP) at Paris-Dauphine University, Jean-Marie Chevalier and Patrice Geoffron; and the directors of the Europlace Finance Institute Foundation (IEF), Arnaud de Bresson and Jean-Michel Beacco.

I have been fortunate in benefiting from discussion with numerous colleagues, whom I thank for their encouragement: Dominique Bureau, Jean-Christophe Bureau, Frank Convery, Patrick Criqui, Denny Ellerman, Pierre-Noël Giraud, Olivier Godard, Christian Gollier, Stéphane Hallegatte, Jean-Charles Hourcade, Pierre-André Jouvet, Jan Horst Keppler, Delphine Lautier, Franck Lecocq, Jean-Hervé Lorenzi, Jacques Percebois, Jean-Pierre Ponssard, Gilles Rotillon, Jean Tirole, Bertrand Villeneuve, Jacques Weber and Jonathan Wiener.

I owe an immense debt of gratitude to the Climate Economics Chair research team; and to CDC-Climat Recherche, the carbon finance subsidiary of the Caisse des Dépôts. I warmly thank all their members. In particular I would like to thank Emilie Alberola, Valentin Bellassen, Malika Boumaza, Stéphane Buttigieg, Henri Casella, Agnès Cassagne, Anaïs Delbosc, Jérémy Elbeze, Damien Fessler, Ignacio Figueras, Natalie Frank, Jessica Lecolas, Stephen Lecourt, Benoît Leguet, Alexia Leseur, Maria Mansanet, Suzanne Shaw, Boris Solier and Raphaël Trotignon.

I would like to thank all those who helped me by reading and commenting on the manuscript, in particular Guillaume Bouculat, Clément Chesnot, Ariane de Dominicis, Jean Jouzel, David Molho and Jacques Saint-Marc.

My thanks also go to my editors Gaëlle Jullien-Picard at Pearson Editions and Chris Harrison at Cambridge University Press, for their informed advice; to Michael Westlake, who was responsible for the translation; and to Gildas Bonnel and Sébastien Louveau, of Sidiese, for their sales promotion of the book.

Naturally, writing such a book made considerable demands on my family – first and foremost I give my fond thanks to my wife Brigitte, for her patience and forbearance, as well as to our children and grandchildren.

Manaus opera house

The first glimmer of dawn lights up the dark water that extends as far as the eye can see. The air is heavy with humidity. The silhouette of a ship anchored in the river stands out clearly. It is an oil tanker from the Urucu oilfield, 650 kilometres upstream, here to supply the refinery in the free-trade zone, where the leading companies of the Brazilian electronics industry are booming. The river is already bustling with local traffic. Small boats are unloading cargoes of bananas and manioc destined for the central marketplace, the metal arches of which extend along the river bank. Street vendors are beginning to spread out their merchandise on the pavement. The opera house is perched half way up the hill, its dome towering above the city. In a few hours' time, visitors will be crowding inside, admiring the stage and the gilt décor, comparable to anything the Paris opera has to offer.

We are in Manaus. Two million people live here, in the heart of the Amazonian rainforest, 1,500 kilometres from the mouth of the Amazon. In 1669, the Portuguese settlers established a trading post to signal their presence to the overly close Dutch explorers in Surinam. The town grew by leaps and bounds in the late nineteenth century. From 1890 to 1910 Manaus was the richest city in Latin America and one of the most prosperous anywhere in the world.

Manaus's wealth was based on the rubber boom. The Amazonian hevea tree, naturally present in the forest, was for several decades the world's only source of latex. Its exploitation developed at a frenzied pace. The rubber barons sent their workforce, the *seringeros*, ever deeper into the forest to tap the hevea trunks and bring back the precious sap. The economic model was straightforward. On the supply side, it was a matter of taking the largest possible amount of latex from a seemingly inexhaustible reservoir: from Manaus, the Amazonian forest is a green ocean extending interminably in all directions. On the demand side, the need for rubber also seemed to be limitless, for competition was nonexistent. Prices, therefore, could only continue rising. The rubber barons' investments quickly took an extravagant – as testified by the town's central marketplace and opera house – and dangerously speculative turn.

With an economy of predation on the supply side and an economy of speculation on the demand side, it was a state of affairs that was bound to come to grief. The downfall of Manaus resulted from what some people view as one of the first acts of 'bio-piracy'. In 1876, the explorer Henry Wickham embarked for London with a collection of 70,000 hevea seeds in the hold, destined for the great greenhouses of Kew Botanical Gardens. Kew's biologists managed to keep alive fewer than 3,000 sprouts from the seeds, which were then sent to Colombo botanical garden in Sri Lanka. Colombo became the hub for hevea seeds destined for Asian commercial plantations, where production started around the turn of the century. The Manaus rubber barons never recovered. The most entrepreneurial of them tried to convert their mining-type production model into commercial plantations. But confronted by the green Amazonian ocean their efforts were in vain and none of them managed to curb the attacks of parasites, particularly microcyclus, that destroyed the young shoots. The decline of Manaus was unremitting and continued until around 1960. Travellers told of the dilapidation of the opera, its structure rotting away in the moist tropical heat. Eventually it had to be closed to the public.

The economic and financial crisis that struck the world economy in 2008–9 has certain similarities with the epoch of the rubber barons. We find the same two ingredients: a speculation economy and a predation economy.

Firstly, take a speculation economy. The bankruptcy of Lehman Brothers, one of the flagships of Wall Street, in September 2008, revealed the fragility of the sophisticated international financial products that were supposed to ensure better risk management by distributing it, through securitization, among a large number of actors. When speculation enters the picture, pushing asset prices ever higher on financial markets, the financial sphere becomes disconnected from economic reality. The prices posted in these markets no longer reflect the real risks to which savers and investors are exposed.

Secondly, take a predation economy. The recession was also triggered by the unprecedented escalation of energy and agricultural prices that accompanied the speculative bubble. This market tension resulted from the spread around the world of a commodity-intensive development model that exerts growing pressure on energy and agricultural resources. But with oil priced at $140 a barrel (in July 2008) and wheat at $450 a tonne (in April 2008), how can the continuity of energy and food supplies be assured? The answer is that it cannot, and this is precisely what induced the fall in world demand that led to the economic recession and the collapse in all commodities and financial asset prices.

Are our societies consequently doomed to the fate of the rubber barons?

The great crises of capitalism mark historical turning points. The depression of the 1930s led to a break with free-market principles by strengthening the role of the state through Keynesian policies of acting on demand and through regulating the financial system. This type of system flourished in the post-war period, 'les trente glorieuses' as the French call it, up until the two oil shocks when the economy sunk into a second major recession. This new crisis resulted in the pendulum swinging back towards the market, championed by Ronald Reagan and Margaret Thatcher, with the opening up of national economies to the free movement of capital. A new era of growth then started, stimulated by the rise in power of the emerging economies, which were swift to join the onward march of globalized capitalism. Millions of people in the emerging world were able to escape poverty and the middle class expanded, aspiring to the same living standards as their peers in the industrialized countries. With this latest crisis, the historical epoch of financial globalization has reached its limits. We have arrived at a new historic turning point for capitalism. The thesis of this book is that climate change will play a major part in what comes next.

Over the past few years, the issue of climate change has had wide media coverage, contributing to a collective awareness of the climate risk arising from the accumulation of our greenhouse gas emissions in the atmosphere. It is nevertheless difficult for the media to avoid being a mouthpiece for two opposed – and mistaken – stereotypical reactions in the face of climate change: catastrophism and climate scepticism.

Catastrophism is used, to a greater or lesser extent consciously, by some proponents of large-scale action to reduce greenhouse gas emissions. It sets out from the fact that past trends are not sustainable and are leading to major disruptions that could even result in the extinction of the human species. This anxiety-provoking approach would mobilize support if fear actually led to action. But in our society where the boundary between the virtual and real worlds is sometimes hard to discern, it quickly becomes counter-productive. In the digital era, the latest disaster film about climate change is no more an incentive to action than the release of the most recent *Star Wars* DVD.

Another undesirable effect of catastrophism is that it paves the way for the climate sceptics. The work of climatologists shows how difficult it is to forecast the reactions of the climate system to the accumulation of greenhouse gas emissions. This difficulty stems from the many uncertainties that enter into the work of researchers and scientific debate. The climate sceptics would like to conduct this debate on the media stage, rather as if one was to put Pythagoras's theorem to the vote on a TV show. All this would be of little significance if the climate sceptics' line did not have a harmful effect. It is sowing doubt in many people's minds and among decision-makers, for whom it is always tempting to postpone action in the name of uncertainty.

Between catastrophism and climate scepticism lies the responsible approach, and it is this that the present work attempts to adopt. Climate change is here treated as a risk. We ask what is the right way to take account of this risk in economic life.

To come up with an answer, we first draw upon the study of economic instruments that have already begun being constructed. At an international level, climate negotiations began in 1992, at the Rio de Janeiro Earth Summit. It gathered pace with the signing of the Kyoto Protocol in 1997 and its coming into force in 2005, which introduced new economic instruments. In that context, in 2005 Europe launched a carbon trading market, which puts a price on carbon dioxide (CO_2) emitted into the atmosphere. The Copenhagen Conference of December 2009 did not bring the progress that many people were expecting in terms of commitments from the large emitter countries. But it opened up new roads for possible cooperation among countries that will be followed if the appropriate economic instruments are set up.

The book also draws on the work of economists who have addressed the subject before us. Such work raises the question of discount rates. Action against climate change generates short-term costs, while its benefits stretch out over the very long term. The use of a discount rate giving a future price to its equivalent today enables us to compare the cost and the benefit that are so separated in time. The lower the discount rate, the more economic calculation demands action be taken quickly, and vice versa. But how do we set this rate? The role of the economist is to clarify this choice that citizens and their political representatives must make in agreeing on a rate that reflects the real importance attached to future generations. This book has no new light to shed on this question. It assumes that climate change will lead our societies to re-evaluate the importance of the long term and explores ways in which effective action can be taken.

Chapter 1 describes climate risk and shows why uncertainty, which is central to scientists' projections, cannot be grounds for inaction. Chapter 2 analyzes how the economic cards will be re-dealt as a result of climate change. It identifies areas of great vulnerability, calling for rapid international action to strengthen the adaptation capacities of the societies concerned.

What is at stake in the energy system is examined in Chapters 3 and 4. This system will have to adapt to the growing scarcity of fossil fuels beneath the ground. But there is more carbon in these deposits than the atmosphere can absorb without risk to the climate. A further constraint will therefore have to be introduced to protect the atmosphere, by capping the emissions resulting from the burning of fossil fuels. Such a cap enables a price to be allocated to emissions, as is shown by the European Union emissions trading scheme. This prototype could serve as the model for a world carbon market.

Chapters 5 and 6 look at the reorientation needed in agricultural and forest resources, which account for a third of the world's greenhouse gas emissions. This reorientation coincides with a deterioration in the global food situation. It will take different routes from those used for energy, by using project-based mechanisms, financed through the emerging price of carbon. The Kyoto Protocol's flexible mechanisms offer a prototype of what such a system might be in the future. The first large-scale test could be the Avoided Deforestation Financing Mechanism that has been discussed at an international level.

The concept of carbon rent introduced in Chapter 7 allows the macroeconomic impacts of the carbon price to be better understood. Carbon rent, artificially created by mankind, imposes a quantitative constraint on the use of the atmosphere, but not on the development of the economy. If well managed, it can shift resources towards actors who are most in need of them and could breathe new life into the economy through the triggering of low-carbon investment. Widening this rent implies growing cooperation among nations. Chapter 8, devoted to international climate negotiation, explores the conditions for its emergence.

ONE

Climate risk

November 2006, Moscow. The temperature is unusually mild. Animal species, governed by the rhythm of the seasons, are unsettled. The bears in the zoo are having trouble hibernating and need to be given additional food. Several species of migratory birds, which have left on their annual journey, are turning back. Disoriented ducks are engaging in courtship displays.

April 2007, France. With temperatures 4.7°C higher than the thirty-year average, it is the warmest April ever recorded by Météo France. Crops are particularly advanced for the time of year, up to three weeks early for grains. Strawberries and cherries from south-west France are already on the market. This earliness is no one-off occurrence. Since 1945, grape harvest dates have advanced by three weeks, even a month, in certain regions. As the French historian Emmanuel Le Roy Ladurie makes clear, variations in harvest dates, recorded in parish registers since the Middle Ages, are reliable evidence for historical fluctuations in temperature.

December 2007, New South Wales, Australia. The small town of Deniliquin invested in the largest rice production programme in the southern hemisphere. In its heyday, it was able to meet the needs of 20 million people around the world. But after five consecutive years of drought, the factory has recently closed. Australian farmers have abandoned rice growing, with its heavy demand for water. The country's exports have fallen, which has contributed to food riots in Africa and the Caribbean. Of course, this is not the first time Australia has experienced drought. Older generations still pass down memories of the drought of 1902, which decimated the sheep and cattle population. Yet the succession of six exceptionally hot and dry years exactly conforms to the climate change scenarios described by scientists. Or is it simply a coincidence?

The Earth and its climate moods

All the indicators point in the same direction: for a century, our planet has been getting warmer. The measurements have shown an increase in average temperature of around 0.8°C since the end of the nineteenth century. This warming has not been continuous throughout the twentieth century. Between 1940 and 1970, the average temperature actually fell, which made some people think that the Earth was heading towards a new Ice Age. From 1970, warming restarted. It was more rapid than early in the century, with the temperature increasing by more than half a degree between 1970 and 2005. The current rate of increase is around 0.2°C per decade. If this rate were to continue, the terrestrial atmosphere would heat up by 2°C in the course of the twenty-first century alone.

Warming is uneven. It is more pronounced at high latitudes than near to the equator and more pronounced over land than over the oceans. The largest increase in temperature, of around 2°C over a century, has been observed in the extreme north-west of the American continent and in certain regions of central Asia. Conversely, a cooling of some tenths of a degree has occurred in certain regions.

The widespread melting of mountain glaciers and of the polar ice caps results directly from warming. Such melting has contributed to the rise in the average sea level, nearly 10 centimetres since 1960.

This rise is also attributable to the expansion of water due to the increase in average temperatures. The rise in sea levels has accelerated over the past twenty years, probably because of more rapid melting than forecasted of Greenland's continental glaciers and of some parts of Antarctica.

The heightened intensity of droughts in certain regions and the greater frequency of floods in others stems from the warming of the planet. This warming has changed the distribution of precipitation and also has an effect on river flows through its impact on glaciers. Other climatic events are modified, such as tropical storms and cyclones, the intensity of which has probably increased as a result of global warming. Finally, certain extreme events, such as tsunamis, are often mistakenly attributed to climate change; in fact, tsunamis are triggered by undersea seismic activity, which is completely independent of the terrestrial temperature. Not all natural catastrophes affecting human society are attributable to global warming.

What is the significance of an average warming of a little less than 1°C in a century? Such a change seems very small compared to the temperature changes that we adapt to from one season to the next or more simply from day to night. We should be wary of this type of comparison, which confuses two distinct disciplines. The short-term variability of temperatures in a given place is the concern of meteorology; their average evolution over time is the concern of climatology. Climatology is a younger discipline, the findings of which have been made known to the public by the Intergovernmental Panel on Climate Change (IPCC), a global network of scientists attached to the United Nations (UN).[1] In France, the work of the IPCC has been popularized by two internationally renowned scientists, Jean Jouzel and Hervé Le Treut.

---

[1] The IPCC was created in 1988, under the aegis of the World Meteorological Organization (WMO) and the United Nations Environment Programme (UNEP).

Since 1988, the IPCC has been examining long-term climate variations. Their findings are conclusive: an increase of nearly 1°C in the average temperature in a century is extremely rapid, and much faster than temperature changes observed over known human history. The reason for this rapid increase is important. For the first time in the Earth's history, the climate is being changed by the activity of a living species – mankind – which has been able to exploit huge amounts of natural resources to its advantage.

On the track of past climates

How can we gain access to reliable data about past climates? The invention of the thermometer by Galileo dates only from the end of the sixteenth century. It was not until nearly three hundred years later, from the mid-1800s, that the first stations systematically measuring temperatures at the surface of the Earth were set up. Since 1979 their data has been complemented by data from satellites, which measure the temperature of the troposphere – that part of the atmosphere closest to the Earth's surface – by means of infrared imaging. It is this system of direct observations that supports the IPCC's diagnosis on global warming in the twentieth century.

Going further back than 1850, climate historians can explore other sources of information to reconstruct the climate puzzle. The archives of previous centuries are one source. In Europe, these enable us to go back to the Middle Ages and provide indirect climatic evidence, such as harvesting dates and observation of glaciers, or even direct testimony on the climate. Historians such as the Englishman H. H. Lamb and the Frenchman Le Roy Ladurie have meticulously scoured these archival accounts. They have matched them with information coming from other evidence of past climates: analysis of the growth rings of trees, or 'dendrology', which allow past climates to be deduced from tree trunks; observation of the traces left on rocks by ice, testifying to the historical advance or retreat of glaciers; and the study of coral, plankton or pollen found in ancient sedimentary layers.

This type of approach enables the climate of the northern hemisphere to be reconstructed back to around the year 1000 AD. It shows that Europe underwent warming that began in around the tenth century in the northern part from Russia to Greenland. This warming then shifted towards the continent and continued until the fourteenth century. Historians refer to this period as the 'Mediaeval Climate Optimum'. It was followed by a period of cooling, the 'Little Ice Age', which probably reached its peak in the seventeenth century and continued until around 1850. During this period, cold and frost, rather than drought, became the main cause of harvest losses and food shortages. The variation in average temperatures between the Mediaeval Climate Optimum and the Little Ice Age was probably no greater than 0.5°C – less than has been observed in Europe over the last thirty-five years.

To go back further in time, researchers use information contained in 'ice cores'. These core samples, consisted of tubes of ice, are taken from Antarctica or Greenland, where the snow never melts. Snow precipitation accumulates over successive winters and is compressed to form ice. The chemical composition of this ice provides a 'history' of the climate. It is possible, from the water molecules making up the ice, to estimate the temperature prevailing when it was formed. Ice cores also contain other information: ash associated with volcanic eruptions, sand carried by desert winds, air bubbles giving information on the composition of the atmosphere in the past, in particular the concentrations of carbon dioxide and methane, both of them greenhouse gases. The most recent drillings allow us to go back 800,000 years, by means of ice cores several kilometres in length.

On the scale of the last million years, our planet has undergone a succession of cold periods, known as 'glaciations', during which a thick layer of ice covered a large part of the globe, and warmer periods, or 'interglacials'. Modern man, who appeared around two million years ago, has experienced several such glaciations, each lasting some tens of thousands of years. Migrating with the advance or retreat of the glaciers, following their herds of reindeer and taking advantage of the passages opened up by the falling sea level, human beings were then able to survive, through their mastery of fire, in permanently frozen zones.

The last glaciation reached its peak around 20,000 years ago. Average temperatures were 5 to 6°C lower than those prevailing now. Ice covered the whole of the British Isles and extended as far south as Lyon. The sea level was much lower and the English Channel could be crossed on foot. Subsequently the temperature progressively rose, and we gradually entered the present interglacial – the Holocene Period – around 8,000 to 10,000 years ago.

The Holocene began in the Neolithic Age, at the end of the Stone Age. It saw the development of human civilization with the Bronze Age, the Iron Age, and the beginnings of agriculture. The Holocene was characterized by great stability of the global climate until the twentieth century. This global stability does not exclude significant regional variations: 6,000 years ago, the Sahara had abundant rain and was home to species such as the elephant and hippopotamus, which cannot survive in today's desert conditions. Over time, human societies were able to adapt to these many changes in the climate system.

The components of the climate system

The sun's radiation is the main source of our planet's heat. When the Earth was formed more than 4 billion years ago, this radiation was 25 per cent lower than the present level. Its intensity increased as the sun aged. But this change is imperceptible on a human time scale: millions of years are needed to make itself felt. On the other hand, the intensity of solar radiation is subject to cyclical variation, revealed by the observation of 'sun spots' at the most intense periods. The number of these sun spots varies throughout the cycle, reflecting 'peaks' and 'troughs' of solar activity lasting several decades. During the Little Ice Age, for example, the number of sun spots observed in each cycle was very low. It is estimated that the sun's radiation fell by 0.2 per cent to 0.3 per cent at this time, which helps explain the Earth's cooling.

The amount of radiation received by the Earth is a function of its average distance from the sun and the inclination of its axis. The Earth's orbit varies under the effect of the attraction of the Moon and of other planets. These orbital changes can now be precisely calculated over several million years. Since the work by the Serbian astronomer Milankovitch, these orbital modifications are considered to be one of the main causes of millennial glaciation cycles: when the average distance between the Earth and the sun increases, the planet cools to a sufficient extent to cause advances in terrestrial ice. Conversely, when the orbit brings the Earth and sun closer together, temperatures rise and the ice retreats. For greater exactitude, we also need to take account of variations in the Earth's inclination, which alters the distribution of the solar radiation received by the two hemispheres.

When it reaches the Earth, solar energy interacts with the atmosphere and the Earth's surface, both of which reflect back part of the energy and absorb the rest. This proportion is another basic component of the climate system.

About a third of the solar energy does not pass through the atmosphere and is directly reflected back into space. A certain number of atmospheric particles, known as 'aerosols', form a screen from which part of the sun's rays rebound. Clouds are the most well-known of these. Other examples of aerosols are fog and smoke, as well as ash expelled during volcanic eruptions, which intensifies atmospheric haze.

When the quantity of aerosols in the atmosphere varies, the temperature is affected – as is clearly evident when clouds obscure the sun. But other perturbations can play a role. The twentieth century's largest volcanic eruption was from Mount Pinatubo, in the Philippines, in 1991. The accumulation of dust resulting from it disturbed the Earth's climate for two years, 1991 and 1992, which were unusually cold. And it was probably the atmospheric accumulation of sulphur-based aerosols, of human origin, that led to a fall in terrestrial temperatures of around 0.1°C between 1940 and 1970.

Once they have penetrated the initial screen of aerosols, the sun's rays then reach the planet's surface. Some parts of the Earth's surface, such as snowfields or deserts, reflect back a high proportion of solar radiation. This reflective property of surfaces is measured in terms of 'albedo', a word deriving from the Latin *albus* (white). Albedo measures the proportion of solar energy reflected by a given surface. It is nearly 90 per cent for fresh snow and 25–30 per cent for sand. By way of comparison, the albedo of a black surface is less than 3 per cent and that of a conifer forest is between 5 and 10 per cent.

The solar energy that is not reflected – about two-thirds of it – is absorbed by the surface and to a lesser extent by the atmosphere. A major physical phenomenon comes into action here: the law of conservation of energy, which states that all energy absorbed by the Earth and the atmosphere must be returned to space in one form or another. Heat is a way of returning this energy. The Earth emits infrared radiation, similarly to a saucepan on a gas ring, where the incoming and outgoing energy are in equilibrium.

If we were to confine ourselves to the parameters looked at so far, this equilibrium would result in an average temperature at the Earth's surface of around minus 19°C. Our planet would be extremely inhospitable. The more clement temperatures that we experience are due to a final natural component of the climate system, the disturbance of which accounts for current warming: the greenhouse effect.

Trapped heat

The atmosphere is a thin film some tens of kilometres thick around the Earth. It is made up of a mixture of gases – air – the composition of which is more or less constant, apart from water vapour which accounts for between 0 and 4 per cent by volume, depending on the prevailing weather and the place concerned. In the case of dry air, the air consists of 99.9 per cent nitrogen, oxygen and argon. The remaining 0.1 per cent contains very small amounts of compounds with the capacity to absorb the infrared rays emitted by the Earth, then to re-emit them in all directions. These compounds are called 'greenhouse gases'. In decreasing order of importance, greenhouse gases consist of water vapour, carbon dioxide, methane, nitrous oxide and several families of fluoride gases such as chlorofluorocarbons (CFCs) and hydrofluorocarbons (HFCs).

Where does the term 'greenhouse gas effect' come from? In a greenhouse, the sun's rays enter through the glass and warm the air inside. This warmer, lighter air rises, through the phenomenon of convection, and mixes with the cooler air above it. But it comes up against the enclosing glass and, as time passes, the interior of the greenhouse becomes increasingly warm. A second phenomenon comes into play in this warming process. The infrared radiation emitted by the ground is absorbed by the glass, which then re-emits some of it into the interior of the greenhouse. This second phenomenon has given its name to the atmospheric mechanism of the greenhouse effect.

Greenhouse gases play a similar role in climate equilibrium. They trap the infrared radiation emitted by the Earth, then redirect it in all directions. Part of this radiation therefore returns to the Earth. This second energy input, arising from the presence of these greenhouse gases in the atmosphere, leads to further heat emission from the Earth. Finally, an energy equilibrium is reached, not at a terrestrial temperature of minus 19°C, but of around 15°C.

The greenhouse effect also occurs on two other planets in the solar system: Mars and Venus. The very thick atmosphere of Venus allows only a small percentage of the sun's rays to pass through it, which would indicate a low temperature. But CO_2, which comprises 96 per cent of the Venusian atmosphere, traps the small amount of radiation that succeeds in passing through the atmosphere. Since Venus is closer to the sun than the Earth, this radiation has twice as much energy as that reaching the Earth. As a result, the average temperature on Venus is more than 400°C – much higher than on Mercury, which is nearer the sun but is not subject to the greenhouse effect.

The mechanism of the greenhouse effect was first described in 1827 by the French mathematician Jean-Baptiste Fourrier. Around 1860, it was documented by the Irish physicist John Tyndall, who studied long-term climate variations. In 1896, the Swedish chemist Svante Arrhenius established the first correlation between variations in the atmospheric concentration of CO_2 and the average temperature of the Earth. In particular he calculated that doubling the concentration of carbon dioxide in the atmosphere would lead to an average temperature rise of around 5 to 6°C. As a resident of Stockholm, where the annual average temperature is 6°C, Arrhenius viewed his findings as good news! His calculations were premonitory.

The warming of the Earth that began in the twentieth century has coincided with an increased greenhouse effect. Around 1850, the concentration of CO_2 in the atmosphere was in the order of 280 parts per million (ppm). In 2005, it reached 380 ppm, a level probably not seen for at least 650,000 years. This concentration is increasing by almost 2 ppm a year. At this rate, the pre-industrial concentration of CO_2 will have doubled before the end of the century. The atmospheric content of methane and nitrous oxide is increasing just as rapidly, as is that of the fluorinated gases, most of which did not exist in a natural state before the development of modern chemistry.

The origin of this change in the atmosphere is not in question. Increasing levels of greenhouse gases are found in the atmosphere quite simply because human beings have put them there. And it will be very difficult to reverse the process quickly: once they are emitted into the atmosphere, greenhouse gases generally remain there for a long time.

Small, but powerful

Apart from water vapour, the atmospheric content of which is not directly affected by human activity, greenhouse gases are relatively stable. Methane, given off by compost heaps or municipal landfills, remains in the atmosphere on average for twelve years. Nitrous dioxide, which is given off by wheat fields after they are treated with nitrogen fertilizer, takes around 115 years to disappear. Fluorinated gases, such as the CFCs emitted in small quantities by air conditioners, refrigerators and freezers, have lifetimes in the atmosphere that can vary from a few dozen to several thousand years.

For CO_2, far and away the greenhouse gas that we emit in the largest quantities, the situation is rather more complicated. We commonly speak of an average lifetime in the atmosphere of around a hundred years for this gas. This is a simplification, for carbon constantly circulates among different reservoirs: the atmosphere, the oceans, the biosphere and sediments.

Carbon is present in the atmosphere in the form of CO_2 and methane. Part of the atmospheric carbon is transformed into the biosphere by plants, through the well-known mechanism of photosynthesis, which converts carbon from the air and forms organic molecules. These molecules are in turn used by other living beings – animals – for food. During combustion or decomposition of organic matter, CO_2 and methane are produced and then return to the atmosphere. Forest fires, respiration and the fermentation of waste are examples of this. Another larger part of atmospheric carbon is absorbed by the oceans, which contain 85 per cent of the existing carbon stock. There it undergoes chemical transformation. Some of the carbon absorbed by the oceans returns to the atmosphere, while some falls to the ocean floor and forms sediments.

Anthropogenic emissions of CO_2 into the atmosphere are a further link in the chain, disrupting the circulation of carbon between the different reservoirs. The burning of fossil fuels – coal, oil, gas – emits

CO_2. Thus a large amount of the carbon stored for millions of years in the Earth's substrata returns, in just a few decades, to the atmosphere and alters the global equilibrium. Deforestation and soil drainage are two further forms of human activity that modify the natural carbon cycle. Not all this carbon remains in the atmosphere, but is distributed among the different 'fast cycle' reservoirs: the atmosphere, the biosphere and the oceans. For this reason it makes little sense to speak of a CO_2 'lifetime' in the atmosphere. On average, however, it requires rather more than a hundred years for a new stable equilibrium to be established between the different carbon reservoirs following the emission of an additional tonne of CO_2 by mankind.

As we shall see below, a tonne of CO_2 has become the benchmark in the economics of climate change. For most of us, this is a somewhat abstract amount. To put it in perspective, in physical terms a tonne of pure CO_2 is a gaseous mass occupying a volume equivalent to that of a large house. Once diluted in the atmosphere at its present concentration, it takes up a volume equivalent to the Grande Arche of La Défense near Paris or half that of the Empire State Building in New York. In terms of emissions, a tonne of CO_2 corresponds to the waste that we produce driving 5,000 kilometres in a medium-sized car or taking a return flight between two European cities 1,800 kilometres apart. A barrel of oil contains 0.43 tonnes of CO_2, and each inhabitant of the planet on average emits 7.5 tonnes a year.[2]

The tap and bath problem

Once they are in the atmosphere, greenhouse gases rapidly disperse, mixing with the other substances making up the air. Where they are emitted, therefore, makes no difference. What affects the climate

---

[2] At the end of the book, the reader will find a series of key orders of magnitude, prepared by Raphaël Trotignon, expressing the data and general concepts used in this chapter in more concrete terms.

is the overall volume of emissions accumulated in the atmosphere, which absorbs the infrared radiation emitted by the Earth. It matters little whether this comes from Europe, Asia, America or anywhere else on the planet.

Because of the long lifetime of greenhouse gases in the atmosphere and the complexity of the carbon cycle, variations in the atmospheric concentration of greenhouse gases are slow to take effect. Thus the concentration of CO_2 in the atmosphere depends little on our current emissions. It results from the accumulation of emissions for more than 150 years by successive generations. Even if we were suddenly to reduce the amount of greenhouse gas emissions worldwide, it would take time for this significantly to influence the level of atmospheric greenhouse gas concentration.

It is very much the atmospheric concentration of greenhouse gases, in other words the amount of these gases present in the atmosphere, that matters for climatologists. Variations in this amount over time are equal to the entry rate minus the exit rate. Entry into the atmosphere comes from emissions. Reductions in the amount result from the absorption of the carbon present in CO_2 by plants and the oceans and from the elimination of methane, nitrous oxide and fluoride gases, after a certain time, through chemical reactions or absorption by solar rays. If we could precisely measure entry rates and their history on the one hand and exit rates on the other, we could then calculate the evolution of the stock. Doing so recalls, in a rather more complicated form, the nightmarish problem of the tap and the bath from school mathematics: how long does it take to fill or empty a bath by means of a tap with a known rate of flow, taking into account leaks from the bath while the tap is turned on? The classic problem of stock and flow.

Thanks to the work of climatologists brought together by the IPCC, we possess orders of magnitude enabling us to link greenhouse gas emission rates and stocks present in the atmosphere.

In 2005, the concentration of the six main greenhouse gases (excluding water vapour) was in the order of 455 ppm CO_2 equivalent.

This concentration includes CO_2 itself, which accounts for 380 ppm, and non-CO_2 gases for the remaining 75 ppm. These non-CO_2 gases are converted into CO_2 equivalent on the basis of their warming power over a hundred years. For example, a tonne of methane in the atmosphere has the same warming power over a hundred years as 25 tonnes of CO_2. It thus counts for 25 tonnes of CO_2 equivalent.

In order to stabilize the atmosphere concentration of the six main greenhouse gases at 455 ppm, we would need to end the observed growth of emissions by 2015 at the latest. It would then require at least halving them by 2050. If reductions in emissions were to be made very rapidly, halving them by 2050 would, under the best hypothesis, be sufficient. If the reversal of the emissions trajectory occurs more slowly, the reduction in emissions needed to stabilize the existing concentration of greenhouse gases in the atmosphere could be as high as 80 per cent in 2050. The more quickly action is taken, the less effort will be needed later.

If no measures are taken, emissions of greenhouse gases of human origin are liable to grow at the same rate: 70 per cent over the last twenty-five years. This estimate is relatively conservative. The rapid growth of the global economy stemming from the economic dynamism of emerging countries has led to the acceleration of emissions during the past decade. Such levels could result in greenhouse gas concentrations in excess of 900 ppm before the end of the century, with hard-to-imagine consequences for the equilibrium of our climate. The planet would need to adapt in less than a hundred years to a greater change in the climate system than that which occurred in the 20,000 years separating us from the last glaciation peak – with major risks of the climate system racing out of control.

Climates of the future

To predict future climates, scientists use models. By means of equations, these models describe the interactions between the atmospheric and ocean masses, the water cycle, the carbon cycle, etc. The equations are then processed by extremely powerful computers. In its fourth report, published in 2007, the IPCC analyzed the results of modelling studies describing the reactions of the climate to no less than 177 emissions evolution scenarios over the twenty-first century.

A first consideration: a very wide range of results emerged from these simulations. On the one hand, this stems from different hypotheses as to the evolution of future emissions. But there is still great uncertainty as to the response of the climate system to the increase in greenhouse gas emissions of human origin. Indeed the climate's sensitivity to the growth of our emissions depends on retroactive effects that are difficult to model.

Some retroactive effects are negative: they cushion the effects of emissions growth. For example, the increase in the atmospheric CO_2 content stimulates photosynthesis; and the more rapid growth of plants enables more carbon to be stored and to 'soak up' part of the additional emissions of human origin. Warming can also, under certain conditions, increase the density of clouds and enhance their property of reflecting back into space part of the energy coming from the sun.

Other retroactive effects are positive: they amplify the warming initially induced by the additional human emissions. The most immediate concerns water vapour, which is the main greenhouse gas: warming increases evaporation and plant transpiration, which in turn augments atmospheric water vapour content and hence the greenhouse effect. The melting of ice and snow reduces the albedo of the surface while increasing its capacity to store energy. The progressive warming of the oceans and of permanently frozen soil, known as 'permafrost', is in the long term liable to release large quantities of carbon dioxide and methane, currently trapped in the sea depths or beneath the frozen soil layer.

Some retroactive effects would have consequences that are rather difficult to anticipate. For example, warming is liable to alter the circulation of the Gulf Stream, which accounts for the temperate climate of western Europe and the Atlantic coast of North America. In the long term, warming could even reverse the Gulf Stream, giving rise to effects like those portrayed in the film *The Day After Tomorrow*: a submerged Manhattan, shrouded in ice, whose population heads south only to encounter heatwave temperatures. Needless to say, climatologists' scenarios do not reproduce those of Hollywood. Yet many indicators suggest that, in the past, abrupt climate changes resulted from alterations in the circulation of the great marine currents. These currents are highly sensitive to the salt content of the ocean surface, which can in turn be affected by the advance or retreat of ice following variations in temperature. We therefore cannot ignore this type of scenario in evaluating the risks arising from climate change.

Uncertainty as to the response of the climate to an increased greenhouse effect is even greater when we descend to a regional level. Climatologists predict that in 2050 Paris will be considerably warmer than it is now. But they are uncertain as to whether its average temperature will be more like that of today's Montpellier or today's Seville.

A second consideration: none of the projections derived from the models leads to a rapid stabilization of the climate in the twenty-first century. In the (hypothetical) case where the atmospheric concentration has been stabilized at 2005 levels, the temperature would continue to rise for several decades and the sea level for several centuries. At the other extreme, scenarios based on a continuation of current emission rates portray a world unquestionably at great risk: with 900 ppm CO_2 equivalent in the atmosphere by the end of the century, there is a high probability that the average rise in temperature over the twenty-first century would exceed 4°C. The rise in sea level could be more than a metre, and this would accelerate after 2100. Extreme climate events would proliferate, and the models have difficulty describing the accumulation of reactions that could result in a runaway climate system.

Between these two extremes, there are many possible combinations of emission scenarios and climate change. All show that if we want to limit the average temperature rise, we must radically reduce anthropogenic greenhouse gas emissions.

A third consideration: the amplification of the greenhouse effect changes not only the average temperature and the rise in sea level that automatically results from this, it also reconfigures all the parameters of the climate system. The level and distribution of precipitation are affected: more rain on average, but less well distributed than today. River flows are altered. Thus it is the whole of the water cycle that will be modified. In addition, melting ice risks disrupting the circulation of ocean currents, which in turn affects the wind system. Exposure to extreme climate events will be more frequent and likely lead to greater damage.

The fourth IPCC report provides very detailed documentation on regional variations in climate change. It shows that different countries will be affected very unequally. In the industrialized countries, the negative impacts will generally be less severe and will appear more gradually than in countries of the South. Because of their geographical exposure, Australia, Japan and the US will be more affected than Europe. Canada and Russia will benefit for several decades from very positive effects from climate change: the extension of agricultural zones northward, the opening up of new maritime routes as a result of the melting ice pack, and easier access to new hydrocarbon deposits.

Conversely, the most severe impacts are expected in the most densely populated regions, situated in the developing world. Small islands and high mountain areas are particularly vulnerable, but these account for only a small proportion of the global population. On the other hand, more than 300 million people live in delta areas, on the extensive alluvial plains formed by the long-term accumulation of sediment at river mouths. The most densely populated deltas are located in Asia. More than 80 million people live on the Ganges delta and the Brahmaputra delta (on which Dhaka, the capital of Bangladesh, stands). There are 100 million people on the Mekong delta and at the mouths of Chinese rivers. The Nile delta supports the highest population density per square kilometre.[3] All these populations are very much exposed to even moderate climate warming.

The Mediterranean region, northern Africa, central Asia, Australia and large areas of Latin America will suffer from a massive increase in water stress: more difficult access to water increases the risk of food insecurity. Indeed, agricultural production in most subtropical regions could quickly suffer the effects of climate change, whereas agriculture in high latitude regions could benefit up to a certain point in time.

What climate do we need to adapt to?

As we saw earlier in the chapter, the current average rate of increase in the sea level is a couple of centimetres a decade. There is practically no chance that the rate will fall during the twenty-first century, given the thermal inertia of the oceans. On the other hand, it could well accelerate, even on the supposition that warming is limited to 2°C. Although not so far discernible for most coastal areas, this rise is having negative effects on small islands and often densely populated delta areas. It is these zones that will suffer major, perhaps irreversible, damage in the absence of large-scale societal adaptation.

If all the Greenland ice were to melt, the sea level would potentially rise by about 7 metres. In such a scenario, New York, London, Shanghai and Mumbai would be under water. The probability of such melting occurring between now and the end of the century is very small. But climatologists warn that with average warming of 2 to 4°C, such total melting could occur in the long term. It would be absurd

---

[3] There are also two highly populated delta areas in the developed countries: the part of Netherlands situated on the Rhine and Meuse estuaries, and New Orleans on the Mississippi delta. The Camargue is the main delta area in France, and is considerably more exposed than the rest of the country to climate change, though it is not densely populated.

to commit expenditure today to protect against such a rise in the sea level. On the other hand, urban planning decisions having an impact on the urban fabric in the long term should take account of these potential impacts of warming.

Let us imagine that all the land-based ice – 80 per cent of it on the Antarctic continent, which is much less exposed to warming than the Arctic Ocean and Greenland – melts. The sea rises by 70 metres and engulfs Paris and most of the world's major cities. We are here in a situation of totally uncontrolled phenomena, which are liable to occur if the average temperature rises more than 4°C in the course of the present century. Conventional adaptation actions become powerless in the face of this type of risk.

The example of the rise in sea level can be generalized. With the fourth IPCC report, we can distinguish three climate risk levels according to the intensity of warming expected.[4] The risk level depend first of all on the trajectory that greenhouse gases follow this century.

If we take the rapid emissions reduction scenario, the average temperature rise in the course of the century can be stabilized at below 2°C. In this case, the impacts that will have to be adapted to are in line with those that have already been observed. The relation between the damage and benefits will vary according to the activities concerned and the region, with the positive or negative balance depending on the effectiveness of the adaptations made by different societies.

If the average warming goes up into the 2 to 4°C range, the impacts will worsen. The beneficial effects of moderate warming, such as raised agricultural productivity in high latitude zones, will reverse. The negative effects which were already hitting certain areas will be amplified. Dangerous dynamic processes, such as hard-to-reverse

---

[4] For more detail, the reader should look at chapter 19 of the IPCC Assessment Report by Working Group II, 'Assessing Key Vulnerabilities and the Risk from Climate Change' (IPCC, 2007: p. 781).

melting of the Greenland ice, become probable. The role of adaptation to climate change then consists primarily of minimizing the costs of these impacts, and its effectiveness decreases.

If no collective action manages to curb greenhouse gas emissions, our societies risk finding themselves confronted at the end of the century by scenarios in which the temperature rise is more than 4°C. Above this threshold the risks of runaway climate change proliferate. Some people conceive of turning to unconventional policies and using geoengineering techniques in order to act directly on the functioning of the climate system and to avoid the worst.

Can geoengineering change the situation?

Faced with the climate challenge, so-called geoengineering techniques are sometimes put forward as a kind of miracle solution. These techniques involve artificially manipulating the climate system to counter the spontaneous reactions to the growing greenhouse effect. They fall into two categories: one, techniques aimed at preventing a proportion of the sun's rays arriving at the Earth's surface and, two, those that involve artificially intervening in the carbon cycle.

From a physical standpoint, the warming effect that results from the increase in the greenhouse effect can be entirely offset by a reduction in the intensity of solar radiation arriving at the Earth's surface. If this intensity could be successfully controlled by placing some kind of screen between the Earth and the sun, it might be possible to regulate the average temperature at the Earth's surface. Such a strategy would not, however, help combat the other negative impacts of the increasing greenhouse effect, such as the acidification of the oceans.

In order to reflect the solar rays before they reach the Earth, we naturally think of putting into orbit satellites with large reflective panels. The most advanced project of this type, supported by the US National Aeronautics and Space Administration (NASA), is however rather different. It would involve sending billions of micropanels well beyond the atmosphere, 1.5 million kilometres from the Earth, where the terrestrial and sun's gravitational pull cancel each other out. These micropanels would enable the sun's radiation to be deflected, so that only part of it would reach the Earth. The advantage of this highly futuristic solution is that the chemical composition of the atmosphere would not be altered.

For geoengineering projects that involve increasing the quantity of aerosol particles in the atmosphere, this is not the case. The creation of artificial clouds would be one possibility, but its effect on the climate is highly uncertain. Putting sulphur into the atmosphere is another response to warming envisaged by Paul Joseph Crutzen, the Dutch chemist who won the Nobel chemistry prize for his work on the ozone layer. His idea is to send up balloons filled with sulphur, which would release their cargo at an altitude of 20 kilometres, thereby creating a layer of aerosol sulphates that would bounce back the sun's radiation. Crutzen does not recommend implementing such experiments on a large scale, but rather carrying out studies to gain greater knowledge about the possible impact of such a release of sulphur into the atmosphere if the situation warranted it.

The other main area of research concerns the artificial modification of the carbon cycle, by removing some of the CO_2 present in the atmosphere and storing it in another reservoir. Carbon capture by pumping CO_2 from the atmosphere into underground reservoirs is one possibility that has been subject to research in the US. If economically viable capture processes were developed, they could become a useful complement to the carbon capture and sequestration techniques for carbon emitted by power stations being tested in Europe and the US. Alternatively, some scientists envisage artificially increasing the carbon storage capacity of the oceans by stimulating plankton growth by means of iron sulphate. But many other scientists are sceptical about the effectiveness of such a process.

No doubt there are many other geoengineering projects being dreamt up. Such projects for direct intervention on the functioning of the climate system have some ardent supporters and many detractors. The truth is that such projects are frightening. The objective – maintaining a climate fit to live in on Earth – is unquestionably laudable. But the risks associated with such interventions have by no means been fully explored and the institutional conditions in which they could be implemented are not clearly established at an international level. Some of these projects may perhaps in the future come to enlarge the portfolio of options available to counter the effects of warming. But it would be unwise to rely on options that may become effective tomorrow to justify inaction today.

Taking account of uncertainty

A twofold uncertainty thus applies to climate risk. The first stems from our limited knowledge of the phenomenon, given the current state of climate science, while the second concerns the difficulty of anticipating what kind of new technologies will be available to future generations confronted with the effects of climate change.

Such uncertainty is sometimes evoked to delay taking action: would it not be better to wait so as to have more solid information as a result of future scientific advances? This argument merits being carefully weighed up.

Uncertainty as to the rate, duration and extent of climate change remains high, but advances in climate science seem to shift the emphasis of uncertainty rather than reduce it. The scientist is rather like the locksmith who is asked to open up the way ahead by cracking a locked door. But each time a door is opened, he finds two more doors with different locks. The advances of climate science are resolving many questions and reducing uncertainty. Simultaneously they give rise to new questions. Uncertainty shifts to new fields opened up by research. The four Assessment Reports published by the IPCC from 1990 to 2007 have enabled many doors to be opened. Changes in the climate are increasingly well observed. The models used are more sophisticated and provide increasingly exact simulations. But no climatologist currently claims to be capable of describing the climate that we will produce in future, for it depends on the total volume of our greenhouse gas emissions.

Seventeen years have gone by between the first and fourth IPCC Assessment Reports. In this time a great deal of new information has been mobilized, which has improved our knowledge of climate risk. This information shows retrospectively that if IPCC diagnoses were inaccurate, it was certainly not for the sake of painting a blacker picture, but rather out of a measure of caution. The changes observed in the climate are in fact more rapid than those expected by IPCC experts: the rate of temperature increase over the last twenty years is at the higher end of the range of their forecasts. The rise in sea level is definitely greater than what was modelled, due to the acceleration in the melting of ice. Furthermore, all the big insurance companies know perfectly well that the frequency and intensity of extreme climate events has increased.

Waiting for scientists to fit together all the pieces of the climate jigsaw puzzle before initiating action could result in a very long wait in view of the amount of information to assemble. Given the risks of runaway climate change, such a delay could put future generations in an impossible situation, even if they were to have the knowledge and the technical means that we have no idea about today. It is in this type of situation that the 'precautionary principle' applies. This principle recommends, in the case of a very serious potential threat, that we should act without waiting to come up with a comprehensive scientific diagnosis.

Should we believe the IPCC (or the climate sceptics)?

Is it reasonable, because of the many uncertainties, to doubt the very existence of climate risk? This is the position taken by the climate sceptics, whose most influential representatives are the bloggers Steve

McIntyre and Anthony Watts at an international level and Claude Allègre in France. Their influence has been strengthened by the criticisms directed at the IPCC in a campaign that the media have dubbed, rather exaggeratedly, 'Climategate' (with reference to the Watergate scandal that led to President Nixon's resignation).

One of the strengths of the American bloggers is to scrutinize all IPCC publications and the studies they draw upon and to track down any mistakes or approximations. For despite every care taken in verification and validation procedures, there are bound to be a number of mistakes in reports that run to several thousand pages. Here are a few examples.

In the fourth IPCC report we find the claim that 55 per cent of Dutch territory lies below sea level, whereas it should be the area liable to flooding. Or that the Himalayan glaciers could vanish by 2035, whereas in the original study the year mentioned was 2350. Finding such mistakes and correcting them is very useful. But this in no way undermines the overall diagnosis of the IPCC assessments.

More seriously, the intellectual integrity of certain climatologists has been called into question by the 'Climategate' polemics. A dossier comprising 4,000 emails and documents from the University of East Anglia climate research unit, which has collaborated with IPCC work, has been posted on the Internet. One of these emails seems to indicate that data on temperature records could have been falsified. The dossier was widely publicized and has fuelled criticisms of IPCC practices and led to accusations that the IPCC doctored figures in order to exaggerate climate risks.

One can understand the appeal for part of the media of this type of story, with its similarities to spy novels. But it is clear that it has little in common with the protocols of scientific debate. Many of the criticisms directed by the climate sceptics at the IPCC have turned out to be ill-founded, such as those concerning temperature records from terrestrial stations and satellites. Other criticisms may point to real problems and usefully serve as inputs to debate among scientists. But the truth of Pythagoras's theorem or the principles of thermodynamics are not debated by all and sundry in the local pub or on TV. In the same way, debate around climate science should take place under the right conditions and well away from prejudice or dogmatism.

As regards science, the expression of doubt plays a vital role and sceptics have a say in the matter. However, in the name of information balance, isolated opinions should not be given the same weight as the conclusions of thousands of scientists over a number of years. Criticizing the IPCC is one thing, disputing climate change on the basis of imperfections that have become apparent in the functioning of a human institution such as the IPCC is something altogether different. There is admittedly much to be done to reduce uncertainty around the IPCC's claims and forecasts and to improve its operating methods and communication. But the often passionately expressed certainties of the IPCC's detractors reflect a frame of mind that is far from the calm discussion needed for good decision-making.

Our responsibility in the face of climate risk

Let us sum up. From its emergence in the Holocene 10,000 years ago, the modern human species developed in a relatively stable interglacial cycle. These climate conditions made possible the birth of our present civilization. Throughout this era, human societies have managed to adapt to many changes in the climate.

The modifications that appeared in the twentieth century are of a different kind. With the accumulation of man-made greenhouse gas emissions, the climate system is being subjected to a shock without precedent since the appearance of hominids two million years ago. Climatologists are unable accurately to forecast its reactions to this shock. But they tell us that the reactions will unfold over several centuries and that they will be on a massive scale. It is, moreover, too late to avoid them.

To act responsibly in the face of climate risk, our societies need to take account correctly of the messages coming from the scientific community. When these messages are uncertain or contradictory, they should be discussed within that community, taking account of the greatest possible diversity of viewpoints. Advances in climate science remove some of these uncertainties, but give rise to others. If we had to wait until scientists resolved all the uncertainties before acting, the initiation of action would be constantly deferred – to the great satisfaction of the climate sceptics.

Climate risk justifies applying the precautionary principle and initiating action immediately. The precautionary principle is related to the concept of prevention used in relation to insurance. The main difference is that it applies to the major risks resulting from the warming of the planet. The action to be taken right now corresponds to an insurance premium that the present generation needs to pay immediately to prevent irreparable damage threatening the generations that will succeed us.

And, as John Houghton (2010) reminds us in his book *Global Warming: The Complete Briefing*, when we take out an insurance policy for our car, it is not so much moderate risks that we want to protect ourselves from, but extreme risks: those that are less predictable but more costly.[5] It would be a terrible lack of collective responsibility if we did not treat climate risk in the same way.

---

[5] John T. Houghton was professor of atmospheric physics at Oxford University before becoming director of the British Meteorological Office and chairing the first IPCC working group. His book is one of the thirty key readings recommended and commented upon at the end of the book.

TWO

Some like it hot:
adaptation to climate change

It was an Englishman, Arnold Lunn, to whom we owe the development of Alpine skiing. Founder of the Alpine Ski Club in 1908, Lunn organized the first skiing competitions in Switzerland and Austria. Through his efforts, the discipline was recognized by the International Ski Federation in 1930, then by the Olympic movement. Nearly 660 skiable areas are currently used in the Alps. But this environment is particularly sensitive to warming: the Alps is one of the parts of Europe where the temperature is rising fastest. A recurrent shortage of snow is already hampering the use of some sixty ski resorts. If the thermometer rises on average by 2°C in the coming decades, a hundred or so further resorts will face a lack of snow. If it goes up by 4°C, barely 200 resorts will still be able to operate. Resorts are spontaneously turning to making artificial snow. This type of adaptation increases energy use, which raises operating costs and emits greenhouse gases. It requires water – in France more than 10 million cubic metres each winter – which is expensive to transport. But there remains one essential ingredient for the snow cannons to function: cold. When the temperature refuses to fall, the cannons cannot be used and skiers have to head up to higher altitude resorts or convert to hiking. The unfortunate Mr Lunn would no doubt be turning in his grave.

In the West, the Maldives are primarily known for the images of atoll paradise promulgated by tour operators. The archipelago is made up of more than a thousand tiny islands, of which 203 are inhabited and 90 converted into hotel-islands for tourists. The resident population is in excess of 300,000, more than 1,000 people per square kilometre. More than half the Maldivians live in the capital, Malé, where buildings rise up vertically. An investment programme, launched in 2001, aims at constructing a wall around the city to protect it against the sea. With the highest point of the archipelago less than three metres above sea level, the fight against tidal waves and the salinization of the water table is age-old. One of the first decisions by President Mohamed Nasheed, elected in November 2008, was to add a new component to the plan adopted by the public authorities to adapt to climate change. He created a sovereign fund, fed by a tax on revenue from tourism, intended to guard against climate risks. Part of the fund is devoted to purchasing land in Sri Lanka and southern India so that the population can migrate in the event of too great a rise in the sea level.

High mountain zones and small islands are environments that are particularly sensitive to climate change. The impact of such change is already making itself felt and people have taken action to adapt. As we advance through the twenty-first century, the various impacts of climate change will increase. All of our economic systems will need to adapt.

Winners and losers

From an economic standpoint, the impact of warming on operating conditions of Alpine ski resorts may be analyzed as a loss for the whole sector. At first, it takes the form of an increase in production costs, and subsequently of a loss of capital with the likely closing of lower and middle altitude resorts. This loss can be reduced if investors correctly anticipate the impacts and sufficiently cut back on their investment so as not to be forced to write off installations that have not yet been fully paid for. The overall loss in the sector will be the result of many adjustments between some actors who will play the game well and others who will be weakened and sometimes doomed to disappear. The rules of the competitive game alter: the future of operators specializing in middle altitude resorts who do not manage to convert is put into question; operators specializing in the highest resorts – generally those with greater means – will on the contrary benefit from new customers (shifting from lower altitude resorts) and relative advantages in terms of costs. For a few decades they will greatly benefit from warming, though with uncertainty as to how long, for in the long term there will be no snow at all and everyone will lose out.

We find a similar situation in agriculture, the sector which, along with forests, will be most affected by future climate change. As long as the average rise in temperature is confined to 2 to 3°C, farmers in high latitude regions will benefit from warming, since cold is the primary factor limiting growing and harvesting. The main beneficiaries will be Russian, Canadian and northern European farmers. There will at the same time be many losers, because of growing difficulty in gaining access to fresh water – the main factor limiting farming in subtropical regions, where very many producers in developing countries are concentrated. In the longer run, if warming exceeds 2 to 3°C, experts expect that the decline in agricultural yields and supplies will also affect the temperate zones. World food security would then become a real problem.

The tourism and transport sectors employ more people in the world than practically any other service industry. They will be greatly affected by climate change. This is clearly the case for tourism in which sun, snow and water availability and safety are major criteria in the choice of destinations. But it is also the case for transport. Warming will facilitate the opening up of regular maritime routes that will save several thousand kilometres between Europe, Asia and North America through the melting of the Arctic ocean ice.

A potential gain for certain countries, mainly Canada and Russia, it will also entail a loss of traffic for others. Still in the domain of maritime transport, a large number of ports will have to adapt their infrastructure to provide storage and handling facilities for merchandise. Such infrastructure, located to the rear of ship docking and loading installations, has generally not been designed to withstand even a moderate rise in sea levels. Further examples could be given of the vulnerability of terrestrial transport infrastructure of various kinds.[1]

The tasks facing managers of energy and water infrastructures will be greatly affected by changes in the climate. Warming will first of all alter seasonal demand for energy, with less need for energy in winter for heating but potentially more in summer if the growing use of air conditioning as a form of adaptation continues. The production and distribution of energy will also be affected, due to the location of the infrastructure: nearly all the main power generating plants are situated near the sea or by rivers, so they can easily make use of the water used for cooling. They are consequently directly exposed to changes in river flow rates, flood risks and rising sea levels. Infrastructure dedicated to the supply of drinking water and to waste water treatment will also be affected by warming, for a simple reason: any changes in the climate will substantially disrupt the water cycle. Here too, not all operators are affected in the same way, and the cartography of vulnerability linked to climate change has become a parameter that plays a part in the competitive game and in investment decisions.

How can the global impacts be assessed?

Summing up all the potential gains and losses resulting from climate changes seems like mission impossible. Since the impacts are different

[1] The economists of the CDC Climat at the Caisse des Dépôts have produced several key studies on the issue of the vulnerability of infrastructure to climate change. See: www.cdcclimat.com.

from one place to another on the globe, we would need to be able to gather together countless empirical observations. The time scale to take into consideration is not obvious either: the time scale of climatologists – decades and centuries – is far longer than that generally used by economists. And how do we take account of uncertainties? As well as the climatological uncertainties discussed in Chapter 1, we also need to include uncertainties as to how economies may react to an exogenous shock that has no real historical equivalent. If economic systems are resilient, they can cushion and absorb the shock. If they are not, they will on the contrary magnify the negative effects. Consider the example of the increased scarcity of water in sub-Saharan Africa: growing water stress can further destabilize agricultural systems that are often already very vulnerable and lead to large-scale migration. But it can also lead to the revitalization of villages through mobilizing investment to develop water resources that are currently under-exploited.

To try and aggregate the various impacts, a number of so-called 'integrated assessment models' have been developed by various teams around the world. It was on the basis of such a model that the British economist Nicholas Stern estimated that climate risk could be expressed as a loss in world gross domestic product (GDP) of between 5 and 20 per cent, in other words a far greater shock than anything seen in peacetime during past economic crises. Another major consideration is that 80 per cent of the costs are likely to be borne by developing countries, in particular because of the great dependence of these economies on agriculture. These figures, widely seized on by the media, should not lead us to imagine that there is consensus among modellers on the subject. For example, William Nordhaus of Yale University, a pioneer in integrated model building, confirms that the adverse effects will be more severe in the developing countries. On the other hand, his absolute assessments are far lower than Stern's.

These models depicting the interactions between climate scenarios and the economic system are useful for testing the impact of shocks or comparing variants. For example, it could be interesting to compare the economic effects of two different climate warming scenarios, one with the Greenland ice sheet melting and the other not. A significant feature then emerges: global impacts are not linear. Above certain thresholds, the cost of climate-related damage can rise very rapidly, because the economic system is destabilized. These thresholds are generally reached more quickly in poor countries having fewer economic resources. Most of the models indicate that it becomes particularly difficult to check this acceleration of the global costs due to climate change above average warming of around 2°C and that we enter the high risk zone above 4°C.

Such models can also test different possible reactions by the economic system. For example, the consequences of warming on the food situation are very different depending on the hypotheses one uses for the functioning of world agricultural product markets.

On the other hand, the use of models to give a figure (or a range of figures) as to the global economic cost of climate change raises more questions than it provides answers, in particular because it is very difficult to anticipate the capacity of economic systems to react to climate change.

Spontaneous adaptations in the face of climate risk

Adapting to temperature variations, rainfall and atmospheric conditions is something we do every day, with greater or lesser success – for example, by taking an umbrella with us on a rainy day or forgetting to do so. Adapting societies to climate change rests on much the same principle, but over the very long term. Adaptation to changes in the climate can thus be defined as a set of organizational, locational and technical developments that societies need to implement in order to limit the negative impacts of these changes and to maximize their beneficial effects.

In some ways, one can foresee that societies which are already able to adapt to the vagaries of the weather will manage to adapt more easily to long-term changes in the climate. Over the centuries, the Netherlands has built a network of protective infrastructure against the sea and an institutional prevention and warning system that has proved itself in the face of extreme events. It can be predicted that the Netherlands will adapt more easily and at a lower cost to rising sea levels than, say, New Orleans, where Hurricane Katrina revealed the city's many weaknesses in the face of a surge in the water level.

A high proportion of the adaptations to climate change will be made in the same way as spontaneous adaptation to meteorological variations. Farmers will gradually change what they grow according to local temperature and rainfall conditions, tour operators will offer new destinations in accordance with changes in sunshine and cloud cover, manufacturers will turn to good account the increase in heat-waves by selling more soft drinks and of the number of rainy days by selling more raincoats and umbrellas. Pioneers in the study of adaptation to climate change, such as the economists Samuel Fankhauser, Robert Mendelsohn and Richard Tol, have emphasized the irreplaceable contribution of market forces to this kind of spontaneous adaptation.

One of the difficulties in measuring the costs and benefits of the impact of climate change lies precisely in the choice of hypothesis adopted as to the type of spontaneous adaptation that actors will practise. The first studies of the costs of climate change were carried out in the US. These showed, for example, that moderate temperature rises would quickly lead to falling agricultural production due to the lack of adaptation by farmers. Later studies introduced more realistic hypotheses as to crops and crop rotation. They showed then that the benefits would outweigh the losses for American agriculture so long as average temperature rises did not exceed 2 to 3°C. Beyond that level, the effects reverse, since the spontaneous capacity of farmers to adapt can no longer benefit from warming.

The loss of effectiveness of adaptation strategies when the impacts of climate change reach a certain threshold is an incontrovertible fact. We have already seen this from examining the effect of climate change on the winter sports industry. We again find it when we analyze the role of insurance in adaptation to climate change.

Most of the risks associated with fluctuations in the weather are priced and mutualized by private insurance companies, which calculate the meteorological risks that they are exposed to from probability tables, and then include them in damage insurance premiums.[2] If we were to rely solely on spontaneous strategies of adaptation to climate change, we would ask insurers to price future climate change costs and incorporate them into insurance premiums so as to encourage agents to adapt better. For example, a house on the Ile de Noirmoutier, an island off the Atlantic coast of France, which is almost as vulnerable as the Maldives to rising sea levels, would have an increasing insurance premium geared to the growing risk over time. Insurance will play, and already plays, an important role in the pricing of risks linked to climate change. However, it cannot take on this task alone, as reinsurers regularly point out.[3] On the one hand, it is difficult to assign probabilities to the risks, since the probabilities cannot be calculated from historical data when it is precisely history that is changing – and 'non-probabilizable' risks are not easy to insure. On the other hand, when certain thresholds are reached, private insurance companies can no longer provide coverage: they would need

[2] In most countries, there are mechanisms for the public authorities to take responsibility for damage arising from exceptional climate events. This, for example, is the case in France with its 'natural disasters' system.
[3] Reinsurers are insurers' insurers. If, following a damage-causing event, for example a hurricane that has caused a great deal of damage to buildings, an insurance company has trouble paying indemnities out of the premiums collected, it can generally turn to a reinsurer through which it is itself insured.

to set premiums totally disconnected from market conditions that no-one could any longer afford.

Beyond certain thresholds, which are very difficult to determine at a micro-economic level, spontaneous adaptation actions become ineffective. At that point various economic activities must cease and the population has to migrate. Can public policies keep these thresholds at bay so as to protect society better against climate risk?

The role of adaptation policies

The majority of spontaneous adaptations are likely to take place *ex post*, after the impacts of climate change have made themselves felt. We then speak of 'reactive adaptation'. In terms of risk, anticipation and, if possible, prevention is more effective than allowing a problem to happen and then having to fix it. Adaptation policies are the first line of defence in combating climate change. They aim to strengthen the resilience of societies through better anticipation of climate risk. They combine three action levers: information, urban and regional planning, and strengthening the weakest capacities.

The first of these may seem trivial, but if the ordinary citizen, the company manager or the local decision-maker do not have access to detailed information about the expected impacts of climate change, they can only adapt after the event, once changes in the climate have occurred. Accessibility to such information requires first of all that the overall climate scenarios be expressed on a local scale. A dissemination system then needs to be constructed so that the information is generally available as a collective good and not held in private hands. Until such time as such a territorial network exists, adaptation strategies remain at best a public awareness-raising exercise around general issues of climate change.

The second lever of adaptation policies concerns land management, which influences the choice of location for populations and economic activities. This component can be correctly implemented only on a scale below that of the national level, namely towns and regions. It consists of taking account of the anticipated impacts of climate change in land management decisions. The choice of public infrastructure here plays a major role, for it affects the long-term presence and location of people and assets. Urbanization choices are also of prime importance in that a growing proportion of the population live in cities. In view of the exposure of agriculture, rural planning and development is also a major component in adapting areas to changes in the climate.

To adapt effectively, the community cannot, however, leave it entirely up to decentralized actions. History and geography have resulted in a highly inequitable distribution of territorial vulnerability. History has given rise to a centralization of economic resources in certain areas and left many others impoverished. Geography also matters: small islands and deltas are particularly exposed to rising sea levels, high mountainous regions to melting glaciers and reduced snow cover, and subtropical zones to decreased rainfall and droughts that threaten food supplies. Without transfers of resources to the most vulnerable and most densely populated areas, adaptation there will quickly become impossible. The inhabitants will migrate, adding to the problems of adaptation in other areas. Consequently efficiency, as much as fairness, requires setting up corrective mechanisms to avert this type of adjustment.

Disseminating information at a regional level

The collective utility of meteorological information justifies setting up public services that constantly disseminate highly detailed information enabling the population and businesses spontaneously to adapt to the many fluctuations in temperature, rainfall and wind conditions. This information is made available place by place throughout the national territory. If you live in Marseilles, what use is it to you knowing that tomorrow the average temperature will be

14°C in France as a whole and that the wind speed will be 20 per cent higher than normal? Not a lot. This is why weather bulletins never give the average temperature.

In adapting to changes in the climate, we are far from having an information system comparable to that provided by the short-term weather forecast. The winter sports operator in Savoie or the government official in the Maldives to whom one transmits thirty-year average snow scenarios or the overall rise in sea level worldwide are like the inhabitant of Marseilles to whom one gives the average temperature forecast for France.[4] They need to have far more detailed information in geographical terms. The Intergovernmental Panel on Climate Change (IPCC) plays a pivotal role internationally in the advancement and dissemination of global knowledge about the impact of climate change. But the uncertainties proliferate once one needs to disaggregate this information at the local scale: as well as uncertainty in relation to the global climate scenario, there is also uncertainty in relation to the local manifestations of this overall scenario.

Suppose, for example, that climatologists tell us that in fifty years' time Paris will have a stabilized climate comparable to that of Seville today. This is quite a hurdle to surmount, since it would require modifying a large proportion of the city's buildings and the entire water supply system and probably preparing to downgrade a large part of the now unnecessary district heating network. But at least there is a clear reference point, and the city's engineers can get to work and the elected representatives can find out how people in Seville live. Unfortunately no climatologist can today say whether Paris's climate in fifty years' time will not be more like that of Montpellier or Fez. Moreover, it is highly unlikely that the Paris climate will have stabilized in fifty

---

[4] Contrary to general belief, the rise in sea level resulting from warming is very different in one place to another, due to differences in salinity and marine currents, parameters that are themselves affected by changes in the climate.

years' time. Stéphane Hallegate, one of the first French economists to have looked into these questions, draws an important lesson from them: because of such uncertainties, the right adaptation strategies are those that guarantee the maximum flexibility over time, leaving options open on the basis that advances in knowledge will in the future provide further information.

In many cases, adaptation strategies will require knowing the probability of the occurrence of extreme events, which cannot be read off directly from averages. Let us consider two examples: risks of heatwaves and floods.

The heatwave that hit western Europe in 2003 resulted in the premature death of 35,000 people. From 1 August through to 13 August, the temperature was uninterruptedly 7°C higher than the seasonal norm. It was during this exceptional period that the greatest number of deaths was seen. According to climatologists, this type of event is an anomaly with very low probability of occurring in the present climate, but which could, in certain scenarios, happen once every couple of years after 2070. It would then become the norm to which we would have to adapt. But in relation to this future norm, what will then be the characteristics of a heatwave and what will be its probability of occurrence? The answer depends on the distribution of temperatures around the mean.

Another example: river flow rates. Climate warming affects flow rates via the run-off of melting glaciers and reduction of snow cover upstream of the river basins they supply and via changes in rainfall. The interplay of these parameters is complex. In France, a warmer climate is likely to lead to a reduction in average flow rates of rivers, particularly during summer low water periods, less so in autumn and early winter. Indeed, meteorologists predict a fall in the number of rainy days. But at the same time they anticipate an increase in the number of days with very heavy rainfall, which is likely to accentuate the variability of flow rates. Overall, we should expect a reduction in average river flow rates, at the same time as a greater number of floods during periods of very high rainfall, which in turn will be more frequent and intense.

The production and dissemination of information on the impacts of climate change is by no means easy. It requires generating a large amount of information, since an adaptation strategy cannot be based upon averages. The procedure is easier in countries already possessing a systematic observation network in regard to climate variables and research teams capable of analyzing the data. This is rarely the case in emerging countries and never in the least advanced countries. The provision of reliable information on local impacts of climate change is therefore a potentially fruitful way of enhancing North–South cooperation.

Infrastructure choices: beware of Maginot Lines!

André Maginot, a French officer wounded at the front in 1914, was appointed Minister of War in 1922. He originated the fortification system built between 1928 and 1938 that bears his name: the Maginot Line, intended to protect France against any risk of invasion by the German armies. By 1936, the Maginot Line extended from the Alps to the Ardennes. Viewed as impregnable by the French high command, it formed a continuous line of fortifications parallel to the German border. We know from history what happened next: the German armies skirted the Maginot Line to the north, crossed Belgium and broke through the Ardennes. It took them only a few days to reach Paris, without encountering any resistance. Most French soldiers, who were notoriously under-equipped, were assigned to the defence of the Maginot Line, with no possibility of moving. Indeed the construction of the Maginot Line had swallowed up a large part of the available credit to the detriment of materials and equipment. Like military strategy, adaptation policy should take care not to make bad infrastructure choices, liable to lead to a massive waste of money. It must avoid the risk of 'maladaptation'.

The Maginot Line typifies a protective infrastructure dedicated to the defence of a territory, as are dykes against a rise in the sea level, drainage systems for dispersing heavy rain, or hedges around fields to diminish wind intensity.[5] The development of such infrastructures must avoid the three mistakes made by the builders of the Maginot Line. The first concerns the location of the line of defence. This spatial dimension is important: the adaptation of countries to climate change calls for careful examination of the location of existing infrastructures and of those we want to construct. The second mistake is to poorly assess the risk on the basis of 'false certainties', such as thinking that the Ardennes forest was impassable. Faced with the many uncertainties of the climatologists, we need constantly to track down false certainties and maintain the maximum number of options for the future. The third mistake, in the case of the Maginot Line, was to underestimate the impact of the infrastructure on the way the social system constituted by the army was organized. To effectively contribute to a country's defence, infrastructure can only be one of the components of a more general human organizational system. It is people, and their capacity to adapt, who defend a country, not blocks of concrete.

Despite the practical difficulties, taking account of climate change in the management of infrastructure cannot be deferred, for a simple reason: their lifetime regularly exceeds ten years for industrial infrastructure, fifty years for buildings and forests, and even longer for transport and water management infrastructures. They must therefore be managed and defined according to the characteristics of tomorrow's climate.

The technical characteristics of infrastructure take into account extreme climate risks: bridges, roads, oil pipelines are all designed to

---

[5] Perennial plant cover is not generally counted as infrastructrure. As regards adapting infrastructure to climate change, it seems desirable to do so, since there is a return on investment for creating plant barriers against climate extremes or protecting existing natural barriers, such as the mangroves that defend people against flooding in tropical areas.

withstand certain wind velocities, temperatures, rainfall, etc. These risks are almost always assessed on the basis of historical data that reflect past climates. If we give any credence to prospective climate scenarios, an elementary precaution is to adapt what we currently construct to the climate conditions that will prevail in 2050 or 2100. Calculation of probabilities from historical tables is then no longer possible. Other methods must be used. Some countries are beginning to do this. To counter the risks of rising sea levels, Canada's Confederation Bridge and the new waste treatment plant of Deer Island, located in the port of Boston, have been raised by a metre. But these two examples are textbook cases. Most research departments continue to define the characteristics of new infrastructure according to past climates. This saves a small amount of money for current investment costs, but what will be the cost for future generations who inherit inappropriately designed projects?

Infrastructure associated with water management will be particularly affected by changes in the climate. Take for example the city of Quito, the capital of Ecuador. Its water supply comes from the glaciers of the Andes cordillera, which are rapidly melting. To respond to the expected decline in flows, the city will need to build a diversion system at an additional cost of 30 per cent over and above the investments already scheduled. In the cities of northern Europe, the expected increase in periods of heavy rainfall will test to the limit waste water collection and treatment systems. Additional installations will be required to protect the population and property bordering the Rhine against the risk of flooding. At the same time, cities on the Mediterranean will confront growing water shortages, which will call for better management of priorities as well as additional investment. Many further examples can be given. The economics of water will be profoundly affected by climate change during the twenty-first century, as too will the management of the corresponding infrastructure.

The adaptation of infrastructure to the impacts of climate change should not be confined to upgrading it category by category. It also concerns the structuring effects of infrastructure systems on land occupation. In the old industrialized countries, people's spatial distribution reflects the pattern of transport, energy, housing and other infrastructures. These countries have abundant technical and financial means to adapt, but their infrastructure systems conceal a great deal of rigidity, as was shown by the inability of the US authorities to react to Hurricane Katrina. Developing countries have fewer resources, but they also have fewer infrastructures in place, and therefore less rigidity. A major issue in adaptation to climate risk concerns the ability to design then put in place infrastructure systems that do not reproduce the inflexibility and vulnerability of the infrastructure of old industrialized countries. This is particularly the case for urban infrastructures, where the way they are organized shapes some of the responses to climate risk that could be enacted in the course of the twenty-first century.

Cities and climate risk

Urbanization is a major dimension in land planning and management. One in every two inhabitants of the planet now lives in a city, as against one in three in 1960. In the rich countries, more than three-quarters of the population are now urban dwellers, and the growth rate of cities has markedly slowed. In developing countries, cities only account for a little more than 45 per cent of the population. Urban growth in these countries is very rapid, for migration from the countryside continues unabated.

Each urban complex has its own characteristics in confronting changes in the climate, due to its location, history and structure. But they nearly all have one thing in common: the majority of cities are near the sea or watercourses – an access to water that historically served as communication and transport routes. Of the seventy conurbations in the world with more than 5 million inhabitants, thirty-two are built by the sea, often at the mouth of a river. Most of the rest lie on large rivers. These cities are as a result frequently exposed to the risk of rising sea levels or changes in the courses of rivers consequent upon global warming. Periods of heatwave in them are amplified by the relative lack of plant cover. Finally, cities have high population densities, ranging from 1,200 per square kilometre in a spread-out city such as Atlanta up to 17,500 in Barcelona, typical of the concentrated European city.

The warning and emergency aid systems in the event of bad weather form part of the collective services provided by cities. A city may choose to adapt passively to changes in the climate, waiting for their impacts to occur. It thus practises spontaneous adaptation, wagering that its existing protection system will be adequate when the time comes. A proactive urban adaptation policy, on the other hand, involves collecting as comprehensive information as possible on future impacts and attempting to guard against them by acting in advance. To increase its level of protection, it can make use of three types of parameter: the organization of housing conditions, the greater or lesser quality of which ensure differing degrees of protection to the inhabitants; municipal infrastructures, particularly those pertaining to water, energy and public transport; and urbanization patterns, with the land occupancy arrangement and the type of equilibrium aimed for between individual and collective housing and between residential, business and leisure zones.

These three types of parameter greatly influence the life of the city's inhabitants. Not only do they affect people's capacity to adapt to future climates, they also prefigure their future greenhouse gas emissions, stemming from the greater or lesser population density of the city, the organization of transport, and the quality of buildings and the form of their energy supply. At a city level, the action levers as regards climate change concern both adaptation to such change and the emissions levels of the inhabitants – thus making the city a very pertinent site for intervention in the face of climate change.

Such awareness of the role of the city is quite recent. It is reflected by the emergence of various national and international networks of cities, which act in concert to take on public commitments. Cities have also made their appearance, alongside nation states, in international climate negotiations. But in moving from lobbying or declarations of intent to action, cities sorely lack economic instruments. The next stages of their commitment will entail developing such instruments and testing them in practice.

The stakes are particularly high for the cities of the global South. Their administrators, generally under great pressure from their rate of expansion, struggle to install even minimal infrastructure. Much of the urbanization takes place in an unplanned manner, in living conditions of all kinds organized on the margins of planned urban development, usually without access to basic services. The head-on battle against these forms of makeshift shanty towns by means of bulldozers, driving new arrivals still further from the city periphery, is lost in advance: there will never be enough vehicles and police when faced with the number of rural inhabitants attracted by the city. Consequently cities need to be built that are able to grow rapidly and adapt to all kinds of unforeseen circumstances. This is impossible if the aim is to reconstruct the sophisticated infrastructures, with all their rigidities, of the cities of industrialized countries. New forms of organization need to be invented, initially lighter in terms of collective infrastructures, but capable of evolving and adapting over time. The cities of tomorrow are likely to be invented through this battle against the clock. Such, anyway, is the position taken by Pierre Noël Giraud, a French economist who has taken a fresh look at the management of cities in the face of climate risk.

The issue of agricultural adaptation in the global South

The counterpoint to the unrestrained growth of the cities of the global South is the persistence of rural poverty. This situation is attributable to the economic fragility of the countries concerned, which are highly dependent on agriculture and livestock. Climate change will weaken their economies still further, even under the most optimistic scenarios, with global warming limited to less than 2°C. The harsher conditions of agricultural production resulting from the growing scarcity of water will affect the vast majority of developing countries, with two points of particular vulnerability: sub-Saharan Africa and central Asia, both of them highly populated zones where food insecurity has been aggravated on a long-term basis by the economic and financial crisis of 2008–9.[6]

Can we wager, in this scenario, on spontaneous adaptation? For centuries, farmers have been doing this. They have accumulated a store of knowledge that enables them to react better than anyone to climate changes. In many instances, the best that governments can do to help farmers adapt successfully is to make available the most reliable information possible on climate variables and to avoid interfering in their decisions. But the advantages of spontaneous adaptation can hardly be obtained in rural parts of those developing countries that are most vulnerable to warming. These areas suffer from degraded ecosystems in which the traditional balance between people and the natural environment has often been lost. Farmers' spontaneous adaptations – slash-and-burn agriculture, sedentarization of livestock, deforestation – on the contrary increase their vulnerability. To reverse the process, adaptation policies are needed that provide additional resources and encourage the protection of the environment.

Some small progress has been made over the past ten years or so. Since 2001, the least developed countries have been able to obtain multilateral financing to set up National Adaptation Programmes of Action. Such programmes come in the form of projects that these

---

[6] Australia, the Mediterranean region of the European Union and the south of the US will also be adversely affected.

countries decide to implement to strengthen their adaptation capacity. The analysis carried out by Shardul Agrawala and Samuel Fankhauser (2008) on the basis of twenty-two plans that have been submitted shows that more than a third of these projects concern agricultural development. If one includes projects linked to water management, which are often complementary to those targeted at agriculture, the figure rises to 60 per cent of the projects covered in national plans launched by the least developed countries.

But the amounts of money transferred by these mechanisms are pitifully small. The fund engaged in managing these transfers is entirely fed by voluntary contributions from rich countries. It has been able to collect $160 million to finance the plans of twenty-two countries over three to five years. Even if the donors were twice as generous, the amount would still fall far short of what is actually needed.

Because of geography, but also because of the history of development, changes in climate will automatically widen the gap between the food and agricultural production capacities of the North and South. To prevent this happening, there must be investment of resources in these sectors, often largely forgotten by development policies in recent decades, before global warming begins too much to disrupt the agricultural and food-producing systems of the global South. The earlier action is taken, and with means calibrated accordingly, the less it will cost society later to manage future food crises and accelerating migration from the zones affected. Three types of investment seem to have priority in contributing to such action: research and development in genetics in order to adapt the seeds used by farmers to the changes they will face; collectively organizing producers for training, marketing and investment in the means of production; and the creation of suitable rural infrastructures, particularly as regards irrigation and the protection of plant cover.

But large-scale financing of adaptation policies in the South is scarcely possible with the instruments developed so far.

Strengthening and rethinking development aid

Financial instruments to help the global South adapt to changes in climate have until now been largely based on voluntary contributions. The financing of adaptation actions in developing countries passes through four funds attached to the Global Environment Facility. As the name suggests, this institution is not specifically dedicated to climate change, but serves as a financial instrument for various international agreements on the environment: climate, biodiversity and the fight against desertification.

All the financing mobilized up to 2007 was based on voluntary contributions from donor countries. In December 2007, the situation changed somewhat with the launch of the Adaptation Fund at the Bali climate conference. This fund possesses allocated resources that are likely to exceed $1 billion by 2012. While this amount is undeniably an advance on the few hundred million dollars raised so far, it still falls well short of what is needed. Studies published by international organizations on adaptation costs in developing countries are not always based on homogenous data.[7] In particular it is difficult to find out exactly what they cover. But they all give orders of magnitude in tens of billions of dollars a year that must be quickly mobilized to strengthen these countries' capacity to adapt.

Another weakness of the mechanism is the lack of a general investment doctrine based on a definition of priorities, a hierarchy of means to be implemented and a system of indicators enabling the implementation of projects to be tracked. Elaborating such a doctrine requires that the area covered by the adaptation actions and the priority zones and sectors of intervention be defined more rigorously. In short, the climate dimension must be integrated into genuine development strategies.

---

[7] Five estimations of these costs have been given by different organizations or authors. There is a critical analysis of them in Agrawala and Fankhauser (2008).

Such integration has started being practised by actors engaged in development. For the last few years the World Bank has reoriented some of its funds towards agriculture and water, partly because of the impacts of climate change. Taking account of these impacts is now explicitly included in the selection and assessment criteria for its projects. Moreover, the World Bank has developed its own research capacities, as is shown by its 2009 annual report on development in the world, titled *Development and Climate Change*.

Climate change is also a dimension that development non-governmental organizations (NGOs) are taking into account. CARE, one of the largest NGOs in terms of the number of projects implemented, has, for example, published a manual aimed at actors in the field to help them respond to the impacts of climate change. Along with other organizations such as Oxfam, these NGOs are now present in the international milieux where these questions are debated, and they emphasize how much the climate issue is intrinsically linked to development worldwide.

Choosing to adapt in advance

From the economic standpoint, climate change appears as a shock that will affect the functioning of our societies. It is not a question of a one-off shock such as that arising from a recession, a natural disaster or an armed conflict, but of cumulative modifications of the hitherto relatively stable parameters that define a climate. Such modifications are likely to increase during the twenty-first century and it is their cumulative effect that will disturb the functioning of the economic system.

It is far from easy to anticipate how the economic system will react to this shock, since there is no historical precedent allowing hypotheses to be based on observations. The models built by economists show nevertheless that the disruptions are happening more quickly and more intensely in those countries lagging behind in their development. They predict that actors' spontaneous adaptations will lead to the relocating of activities and populations, which could, in the long term, significantly redraw the map of the world economy. They also reveal the existence of thresholds above which it becomes very difficult for the global system to absorb climate shock because of the escalation of costs and of ever-changing risks of breakdown: disruption of international trade, uncontrolled migration, the rise of protectionism, armed conflict.

Faced with these risks, the primary economic choice is what type of adaptation our societies should aim for. One option, the 'wait and see' approach, is to do nothing until such time as the impacts materialize before initiating adaptations. Another option, the 'act responsibly' approach, is to prepare for such impacts before they actually happen. It is this latter, proactive, approach that adaptation policies aim to encourage by activating three levers: informing actors, forward planning for cities and regions, and worldwide development by prioritizing agriculture and the availability of water.

These adaptation actions are the first line of defence of climate policies. Since they can only be implemented on a local scale, they play a vital role in raising the awareness of citizens and actors in the field. Their implementation immediately highlights a problem of equity, for the most vulnerable situations match up with those of poverty and exclusion from development.

Anticipative adaptation is but one of the lines of defence to be adopted in making economic choices in the face of climate risk. But by implementing it, we act only on the consequences of climate change. If we strengthen our capacities to adapt without also acting on the causes of warming – greenhouse gas emissions – the effort required for adaptation risks being greater than our implementation capacities. For this reason we have to ask ourselves about the conditions underlying the pursuit of our mode of development – originating some two hundred years ago – and the accumulation of waste greenhouse gases in the atmosphere.

THREE

Building a low-carbon energy future

Liliana lives in the new housing development of Burlington Park in the San Gabriel Valley, eighteen miles east of Los Angeles. Her parents were attracted by the size of the houses and the large gardens. But like many of their fellow residents, they have found out what this way of life entails in their electricity bills and expenditure on petrol. The family budget is squeezed even tighter because of their monthly mortgage repayments on the house.

Liliana's family does not include its greenhouse gas emissions in the household budget. Each year it emits 6 tonnes of carbon dioxide (CO_2) per person from its two cars and 8 tonnes from its electricity consumption. In total the family's carbon budget comes to 20 tonnes of CO_2 for each member of the family. This is the average per person in the US. If all the planet's inhabitants lived like Liliana, world greenhouse gas emissions would be instantly three times as much.

Marjohan is fourteen, the same age as Liliana. He lives in the south-east of the island of Borneo. Like his two brothers, he helps his father grow rice. The family owns 2 hectares situated right below the village. Water is abundant here. They flood their fields in the time-honoured way. The use of fertilizer has revolutionized rice farming and they can now get two crops a year. The needs of the family are largely met, and the village grocer, a relative, sells the surplus.

Marjohan's family does not keep a tally of its carbon budget either. Its emissions are around 2.5 tonnes CO_2 equivalent per person. This is the Indonesian average, excluding deforestation. The family emits less than a tonne of CO_2 per person for transport and the house – one-fifteenth of that of Liliana's family. On the other hand, their rice farming emits methane, which amounts to a fifth of the family's carbon budget.

Liliana's and Marjohan's ways of life are very different from each other. Both, however, contribute to greenhouse gas emissions. The young American consumes a lot of energy in her day-to-day life. The main sources of her greenhouse gas emissions are the coal used to generate electricity and the petrol used to run the family's cars.

For its energy needs Liliana's family uses more than 7.5 tonnes of oil equivalent per person. This is what each inhabitant of the US consumes on average, and is four times the world average and fifteen times what an inhabitant of sub-Saharan Africa consumes. This energy-guzzling model, propped up by oil and coal, creates recurrent problems for the world's number one economy. It cannot be generally extended to countries that are rapidly developing due to the limits on physical resources. There are many reasons to reconfigure this model from now on so as to prepare for a 'post-oil' economy. Taking climate risk into account should help accelerate this shift.

The human sources of greenhouse gas emissions

Let us jump from the individual level to that of the planet. World greenhouse gas emissions amount to around 45 billion tonnes CO_2 equivalent, or nearly 7 tonnes per person.[1] These emissions can be

[1] The statistical information system on greenhouse gas emissions conceals large grey areas. The two most reliable estimates in recent years were made by the fourth Intergovernmental Panel on Climate Change (IPCC) assessment and by

divided into two main areas: agriculture and forestry, accounting for a third; and energy for a little less than two-thirds.[2]

Emissions coming from agriculture and forestry are hard to measure and therefore subject to considerable statistical uncertainty. They represent between 2 and 2.5 tonnes CO_2 equivalent per person. Agriculture is the primary source of man-made emissions of the two main greenhouse gases other than CO_2: methane, deriving from the raising of ruminants and rice growing; and nitrous oxide, most of which comes from the use of nitrogenous fertilizers. Moreover, the extension of agriculture is the main cause of tropical deforestation and soil drainage, and every year it emits more CO_2 into the atmosphere than the entire world's cars, ships and aircraft. End-of-cycle decomposition of organic matter also emits significant quantities of methane.

Rather less than two-thirds of world emissions come from energy, which represents more than 4 tonnes CO_2 equivalent per person. To a very large extent, these emissions take the form of CO_2 given off during the combustion of fossil fuels.[3] The energy system also emits

the International Energy Agency (IEA). The IPCC came up with the figure of 49 billion tonnes CO_2 equivalent for 2004. The IEA put the figure for 2005 at 45 billion tonnes CO_2 equivalent, a figure that is in accordance with the World Resources Institute's database. The main divergences in the estimates result from uncertainty of the emissions from agriculture and forestry. Moreover these emissions continued growing between 2005 and 2007, but fell between 2007 and 2009 due to the economic and financial crisis.

[2] As well as these two major human sources of greenhouse gas emissions there are also, in considerably smaller amounts, emissions linked to industrial processes such as cement production which uses clinker obtained by a chemical reaction that liberates CO_2, the manufacture of nitric acid, adipic acid and aluminium, and the emission of fluoride gases of which the main source is the use of air-conditioners.

[3] The CO_2 given off by the combustion of biomass is not included in this figure. When we burn renewable biomass, we restore to the atmosphere only that CO_2 which was absorbed by photosynthesis during the plant's growth. The balance sheet is therefore neutral. If we burn biomass that is not renewable, the negative balance is included in the emissions linked to the agricultural and forest system. It does not appear in energy emissions.

methane (from the oil and gas industry and coal mines) and a small amount of nitrous oxide. Emissions arising from energy production are much better understood that those from the agriculture and forestry system.

Invisible and odourless, carbon dioxide is not detectable by the human senses. It is also difficult to measure as it emerges from a chimney or an internal combustion engine. How, then, do we calculate its emissions? The answer is simple: if we can measure the amount of carbon present in the fuel used, we can calculate with great accuracy the CO_2 emissions caused by its combustion. The carbon content is easy to calculate. Though it is difficult to measure the CO_2 emitted through a factory chimney, calculating the quantity of carbon present in the tank where the fuel is stored is straightforward. Cars sold in Europe, for example, are required by law to specify their CO_2 emission per kilometre. This figure is calculated directly from their consumption of fuel, the carbon content of which is known. Similarly for an industry or country, energy-related CO_2 emissions are calculated from the quantities of fuel used. These quantities are generally well known, except in cases of tax or customs fraud.

In order of size, to obtain the energy produced from a tonne of oil, 4 tonnes of CO_2 are emitted with coal, 3.1 tonnes with oil and 2.3 tonnes with natural gas. These standard coefficients, termed 'emission factors', must then be adjusted according to the precise qualities of the fuel concerned. For example, if lignite is used in a power station it emits more CO_2 than standard coal, and the petrol used in cars emits slightly more than diesel.

Four-fifths of the planet's energy needs are covered by fossil fuels – coal, oil and natural gas. Coal is the main source used to generate electricity. Oil is the basis for practically all fuels used in land vehicles, ships and aircraft. The heating of buildings and the power used in factories is largely derived from fossil fuels. In the current state of technology, CO_2 is emitted into the atmosphere when fossil fuel is burnt. In 2006, coal and oil each contributed around 40 per cent to world emissions of energy CO_2 (42 per cent for coal and 38 per cent for oil). More energy was produced from oil, but coal emits more CO_2 per unit of energy produced. Gas represented 20 per cent of energy-related emissions.

Electricity generation accounts for a quarter of world greenhouse gas emissions. It is the primary identified source and the proportion will keep on rising with time, due largely to the growth of coal-fired thermal power stations in emerging countries. Transport contributes 15 per cent of world emissions and its share is also increasing due to the growing mobility of people (using cars and planes) and merchandise (using lorries and ships). Industry and buildings are the two other main sources of energy emissions. These sectors are direct emitters when they burn heating oil or gas in boilers. But a growing proportion of their energy supply is provided by electricity, which does not emit any greenhouse gases at the time of its use. But they are indirect emitters if the electricity they consume has been produced from fossil fuel.

Changes in the energy sector

Until about 1750, the energy system was part and parcel of the agri-cultural–forestry system. If we neglect the contribution of mills and a few easily accessible seams of coal, our ancestors used a single source of energy: biomass. They used it directly by burning wood or other ligneous materials for heating and cooking. A significant part of their energy also came from using domesticated animals to pull vehicles and carry riders. In this instance, the primary energy resource came from fodder grown to feed these animals. Demographic pressure at this time was much lower. Today the planet supports 50 people per square kilometre; in 1750 there were no more than 5 per square kilometre and in the year zero probably already 2 per square kilometre.[4]

---

[4] For all these long-term changes, the reader can consult the extraordinary panorama presented by Maddison (2001).

The use of biomass by humans nevertheless resulted in deforestation over the centuries, beginning in the Middle Ages in Europe and later in North America, which continued until the middle of the last century. This deforestation was the first visible attack by mankind on the natural carbon cycle.

The industrial revolution changed the course of history, thanks to the pioneers who managed to increase the energy available by turning to extraction of coal from the substratum. The energy system then acquired growing autonomy from the agricultural-forestry system. In 2006, mankind drew no more than 11 per cent of its energy requirements from biomass. For the most part, the people concerned live in developing countries where the biomass resource used is not renewed. Around 6 per cent of energy is derived from a new fuel, uranium, which enables electricity to be generated from nuclear fission. Hydraulic power and other renewables provide 3 per cent. The remaining 80 per cent is supplied by the three fossil fuels, with 34 per cent coming from oil, 26 per cent from coal and 20 per cent from natural gas.

The growth rate of energy supplies declined slightly following the first oil shock, but nonetheless remains higher than the population growth. In 1973 each inhabitant of the planet consumed on average 1.5 tonnes of oil equivalent. By 2006 this had risen to 1.8 tonnes of oil equivalent. By far the largest part of this growth comes from emerging countries, where significant progress has been made in terms of access to energy. Nevertheless, energy inequality is still very pronounced, since the different countries are not all at the same stage in the long-term transformation of the energy system.

The low-income countries comprise a billion inhabitants, most of whom have no access to electricity, gas or oil for their heating and cooking needs. With a little less than a sixth of the world's population, these countries account for only 3 per cent of energy-related CO_2 emissions.

At the other extreme, a billion people live in the high-income countries, which as well as the old industrialized countries also include 'nouveau riche' nations such as Ireland in Europe, South Korea and Singapore in Asia and some of the oil-producing countries of the Middle East. These high-income countries, also with a sixth of the world's population, consume excessive amounts of fossil fuels and are responsible for 47 per cent of energy-related CO_2 emissions. With comparable living standards, per capita emissions in the US resulting from 'the American way of life' are nearly double those of Europe and Japan.

Between the two extremes, medium-income countries account for two-thirds of the Earth's population (rather more than 4 billion people). In particular, they include large emerging countries such as China, India and Brazil, which are rapidly expanding their energy consumption in order to sustain their economic growth. For example, between 2005 and 2007 China was on average bringing a new coal-fired power station into service every two weeks. These countries already produce 50 per cent of global CO_2 emissions, but they have considerable room for progress since they are in transition from the energy models of the poor countries to those of the rich countries. In a country such as India, which has joined the group of medium-income countries thanks to its booming growth, a quarter of the population still has no access to electricity or to modern forms of energy for cooking and heating.

Energy insecurity

Energy insecurity may be defined as a situation where there is a high risk of supply disruptions. It should be said straight away that such insecurity is the primary concern of governments and public opinion, well ahead of CO_2 emissions. A basic consequence follows from this: if we want to implement the reduction of greenhouse gas emissions effectively, we need to find methods that at the same time have a beneficial effect on the security of supply. This lesson applies worldwide, for energy insecurity persists in the rich and transitional countries, though it takes different forms from that prevailing in the poor countries.

The first kind of energy insecurity is the continuing severe energy poverty that deprives a large part of the world's population from access to vital sources of energy. It is estimated that 1.8 billion people do not have access to electricity. More than 2 billion use traditional biomass for their cooking and heating needs through the recovery of wood, ligneous material and waste, generally to the detriment of ecosystems. Apart from its cost to the environment, this form of access to energy has harmful health consequences due to the toxic nature of combustion residues. Like illiteracy or malnutrition, energy insecurity is one of the great obstacles to development.

Are the energy systems of the rich and transitional countries secure? Breakdowns, such as the massive power blackout which totally immobilized New York for three days in August 2003, reveal the great dependence of the developed economies on domestic energy production and distribution systems. Beware of under-investment which weakens these often overly rigid systems!

Energy insecurity also has an external dimension. It can be the result of too great a dependence on a foreign supplier – think of the dominant position of Russia in regard to energy supplies to Ukraine or the Baltic countries. Dependence as regards imported oil, traditionally limited to Europe and Japan, has now extended to the US and the new emerging economies. Recurrent geopolitical tensions arise as a result, as the French economist Jean-Marie Chevalier (2009), a specialist in these issues, points out.

Energy insecurity has been aggravated by the increase in oil prices from 2002 to 2008 that preceded the 2008 financial crisis. This rise also pulled up the price of all other energy materials. The shock severely affected not only the poorest people but also the middle classes. Because of the dependence of their ways of life on fossil energy, any sharp increase in their expenditure has disastrous short-term economic and social consequences if no measures are taken to alleviate these. But suppose governments were capable of establishing effective safety nets to avert these socially undesirable effects. Would not the rise in the price of oil then be the surest way of reorienting the energy system towards low carbon?

The oil price and CO_2 emissions

The idea that high oil prices facilitate the turn towards a low-carbon society is widespread. An increase in the price of oil urges consumers to economize, which in turn reduces waste and contributes to the efficient use of energy. It also accelerates the development of renewable energy sources by making them more price competitive. The rise in the oil price above $80 and then $100 a barrel in 2007 and 2008 triggered an unprecedented wave of investment in green technologies. In the Silicon Valley, green economy start-ups took up the running from older, IT-focused businesses.

For all that, the rise in oil prices is far from having only beneficial effects on climate risk. On the contrary, it leads to an increase in energy-related emissions in the medium and long term.

The first impact of higher oil prices is to increase the use of coal, which is cheaper and very abundant: proven coal reserves are estimated at nearly 200 years at current consumption rates, or five times the proven reserves of oil and three times that of natural gas. Coal is generally very cheap to produce, especially in open-cast mines such as those of Australia and the US state of Wyoming, where it suffices to scrape away the surface soil to extract it. Until now, higher oil prices have mainly led to more coal being used in electric power generation and some basic industries. But with a sustained widening in the difference between the respective prices of oil and coal, it would become profitable to liquefy coal and to transform it into vehicle fuel. Many projects already exist for developing this still experimental 'liquid coal' industry, with obvious risks in terms of emissions: a fuel derived from coal emits more greenhouse gas than either of the two grades of diesel now used in vehicles.

The second pernicious effect of high-priced oil is to prolong its foreseeable use worldwide. The forty years of proven reserves calculated by the International Energy Agency (IEA) in 2008 does not at all mean that in forty years' time oil will have been exhausted. Progress in exploration techniques will result in new deposits being discovered. Only a small fraction of the globe has been systematically prospected. It is likely that new deposits will be exploitable in years to come: deeper under the sea bed; in the Amazonian rainforest and under the ice of the Arctic Ocean; in the oil shales that the Canadians have begun exploiting and with which Venezuela is well supplied. It all depends on the price. Thus a high oil price is likely to tie up billions of dollars and will drain all kinds of unconventional oil and gas, the use of which is disastrous for the environment. Would it not make more sense to use the money to explore other energy sources that do not emit greenhouse gases?

The third and final disadvantage is that a sustained rise in the price of oil increases the financial resources of those actors who are economically and politically most interested in prolonging the 'oil adventure' by increasing the oil rents from which the wealth of the oil majors and state producers originates. This boosts the financial means that the oil companies can allocate to their research and exploration investments. Geographically, it strengthens the position of the oil-producing states – which can only make more problematic their adherence to the objectives of international climate diplomacy.

The business-as-usual scenarios developed by the IEA are based on a real oil price that goes above $100 a barrel, and then rises over time due to the growing marginal cost of extracting oil. The decreasing reserves available in already exploited deposits forces operators to invest in new oil fields where extraction is less costly. What are the impacts of this rise in the oil price on the energy system?

By 2030, three-quarters of additional energy demand will come from emerging countries. The growth in the supply will be able to provide every inhabitant of the planet with the equivalent of 2.1 tonnes of oil in 2030, as against 1.8 tonnes in 2006. Nevertheless, energy insecurity will scarcely be reduced. Oil supplies will be based on growing imports from a small number of high-risk countries. In 2030 more than 2.5 billion people will remain dependent on traditional biomass, including a high proportion of the inhabitants of African countries that will have become big hydrocarbon exporters by 2030.

What will be the effect of these developments on greenhouse gas emissions? In 2030, 80 per cent of energy supplies will come from fossil fuels. The contribution of oil will admittedly have fallen, but the main beneficiary of the decline will be coal, which will join oil as the other main energy source used in the world at the end of this period. On average each inhabitant of the planet will emit 5 tonnes of energy-related CO_2 in 2030, as against a little less than 4 tonnes in 2000. Per capita emissions will have slightly fallen in the developed countries, but will have continued growing in the rest of the world. Even more worrying, the investments made in the energy sector from 2006 to 2030 make it unlikely that there will be any rapid reduction in emissions after 2030. The number of coal-fired thermal power stations not equipped with CO_2 capture systems will have increased over this period. Large investments will have been agreed in order to recover non-conventional or hard-to-access oil. A new industry producing liquid fuels from coal will have been formed. These characteristics of energy generating installations inevitably result in continuing increased emission levels which will reach 6 tonnes per capita in 2050. Clearly such a level places the world on an emissions trajectory with a high risk of temperatures rising more than 4°C during the second part of the century. Such, then, is the 'business as usual' scenario.

Investing today to emit less in 2070?

Energy production is one of the economy's most capital intensive sectors. It ties up a large amount of capital over very long periods. A gas turbine operates on average for around twenty-five years. A nuclear power station produces electricity for forty to fifty years. A large thermal power plant can operate for sixty years and the lifetime of an electricity-generating dam commonly exceeds seventy years. Part of the infrastructure that we will be using to produce and consume energy in 2050 is already in place. Furthermore, there is a time-lag of several years between an investment decision and installations actually coming on-stream. Thus today's choices determine the configuration of energy production in 2070, and even beyond.

This inertia makes any rapid reconfiguration of the system difficult. On the other hand, if we rapidly changed the orientation of investments, the long-term effects would be spectacular. But we must not delay. In the industrialized countries, a substantial proportion of energy installations need replacing by 2020. Not taking advantage of this situation to reduce the proportion of high-emissions installations would cost us dearly later. Developing countries are less constrained by the infrastructure already in place. As we have just seen, they will account for a large part of the additional energy produced over the next fifty years. It is crucial that investment decisions enable them to meet their vast energy needs without generating additional emissions that are dangerous for the equilibrium of the climate.

Can a rapid shift in energy investment policies lead to a more satisfactory CO_2 emissions trajectory? Given the population growth, we need to halve greenhouse gas emissions by 2050 to limit per capita emissions to 2.5 tonnes CO_2 equivalent. To achieve this, we must aim to reduce energy-related emissions to below 2 tonnes per capita as against the 6 tonnes in the business-as-usual scenario.[5]

A reduction from 6 tonnes to 2 tonnes of CO_2 per capita in 2050 can be achieved by combining three types of action. The general

---

[5] It appears impractical to plan for zero emissions for all activities excluding energy. Agriculture and forestry need to deliver high growth in food supplies by 2050, and this will make any reduction in their emissions especially difficult.

implementation of energy efficiency could save 1.6 tonnes per person. Increased use of renewable energy and biomass, mainly as a substitute for coal, would bring an additional saving of around 1.4 tonnes per person. That leaves one tonne to be found. This can be obtained by the general adoption of carbon capture and storage in electricity generation and heavy industry and, to a lesser extent, renewed investment in nuclear energy.

In the current economic and political conditions, most of these investments are not profitable and have no chance of being implemented on a large enough scale or within the requisite time frame. Consequently the reshaping of the energy landscape requires introducing new incentives. Extending the role of carbon prices is an important aspect of this, but it is by no means the only one.

The introduction of a carbon price

To create a low-carbon energy system, the orientation of investment needs to be radically altered. The process is unlikely to be implemented evenly. There will be technological and organizational breaks that we do not foresee today. In particular, it is not enough simply to produce energy; it also has to be distributed, and therefore the existing grids carrying power generated from fossil fuels, which are often barriers to the entry of new energies, must also be transformed. How much will all this cost?

At first sight, the bill looks steep. This is generally the case when it comes to the environment: *ex ante*, the costs of environmental initiatives are usually overestimated. There are two reasons for this. It is in the interest of the actors causing the pollution to inflate these costs. But a large part of the expected costs is in our minds, as we are unconsciously conditioned by the current norms of society. We tend to view any radical change as impossible or very costly. Moving to a low carbon energy system is no exception to this tendency.

The various ways of reducing emissions cover a wide spectrum of costs. Many of the actions we are asked to take – adjusting the heating, driving more slowly, switching off the lights – can immediately reduce the emissions burden. Here we are in the so-called 'negative costs' zone, that of actions that can yield a few dozen euros per tonne of CO_2 saved. Buying a hybrid vehicle or switching to a solar heating system represents an over-investment, the cost of which may be partly or fully amortized subsequently, though not necessarily so. In current conditions, it could cost a few dozen euros per tonne of CO_2 eliminated. Travelling by plane without emitting any greenhouse gases is today impossible. Impossible actions are those with an infinite cost.

In the short term, the range of emissions reduction costs thus starts with negative values and tends towards infinity. The notion of average cost therefore makes little sense. On the other hand, it is relevant to arrange reduction actions in ascending order of cost. We thereby obtain the marginal cost curve of emissions, which will increase as the easiest actions to take are progressively implemented. In the short term, marginal costs are low or negative for the first emissions reduction actions. They then increase rapidly. The harder reductions are to achieve, the steeper the slope of the marginal cost curve.

With the passage of time, the constraints shift. New technologies become available. Their cost falls as they spread into the economy, often more quickly than is expected. Progress in providing people with information or in how they are organized can play a similar role in lowering costs over time. What was previously an intangible limit becomes feasible in the long term. The emissions reduction marginal cost curve becomes flatter as time passes. This flattening can be amplified by appropriate policies in terms of technological and organizational innovations.

If we project ourselves forwards to 2050 with today's marginal cost curve, the investment required to go below 2 tonnes of energy-related CO_2 per capita seems exorbitant. But if we assume that this flattening of costs over time occurs, the investment needed becomes more reasonable. Within its long-term technology perspectives, the IEA counts, for example, on an average cost of around $75 per tonne of CO_2 to attain this objective, with the most costly actions reaching $375. Adjusted to the total increase in wealth expected by 2050, the total amount involved represents less than 1 per cent of global GDP. But the experts also assume that economic actors systematically choose to reduce less costly emissions by comparing the cost of their emissions with the price of carbon.

For every megawatt hour (MWh) produced from a coal-fired power station, on average 0.9 tonnes of CO_2 is emitted into the atmosphere. If the use of the atmosphere is free, coal-fired power stations are very often competitive in relation to other techniques given the relative prices of different types of energy. If the generating company pays €20 per tonne of CO_2 emitted, a typical price level on the European CO_2 allowances market, this represents an increase in the current cost price of about 30 per cent for coal-fired power stations. The incentive is sufficient to invest in efficiency improvements or alternative supply systems from biomass. At this level, coal-fired power stations would no longer be financially viable. They would need either to shut down and be replaced by low carbon installations, or to make large investments to capture CO_2 at chimney exit points to prevent it entering the atmosphere.

Setting a price on CO_2 emissions thus provides a double incentive in the energy system. From the standpoint of demand, it prices the use of different energies according to their greater or lesser carbon footprints. From the standpoint of supply, it encourages producers to reorient their investments in favour of low-carbon energy infrastructures. Let us begin by looking at the demand side.

Investment in energy efficiency

The capital tied up is not the only element contributing to the inertia of the energy system. Our behaviour is also a major obstacle to the reconfiguration of the energy landscape. The first step to be taken concerns such behaviour. It may be a platitude to repeat it, but investing in energy economies has the greatest impact on emissions. Savings of around 40 per cent of primary energy use throughout the world can be made by 2050 by systematically tracking down and eliminating all energy wastage. It is not a question of going without this 40 per cent by reducing our quality of life, but of obtaining the same services through gains in efficiency.

Such investment is not expensive. On the contrary, it brings a good return! The experts speak in terms of 'negative cost' measures. There are many possibilities open to us: turning down the thermostat on radiators, disconnecting the air-conditioning, driving slowly, etc. When all such actions are added up, the results are impressive. Nevertheless, we should not rely solely on people's voluntary initiatives in setting out in a sustained way on the energy savings road. We must first ensure that quality information is provided to energy users. And above all effective incentives should be introduced. These incentives are three in number: prices, norms, and city and territorial organization.

The price of energy is the first signal. As soon as the price of energy rises, we tend to economize. But this stimulus is reversible, as was seen during the oil counter-shock of the 1980s. In addition, apart from its unpopularity, it also raises questions of fairness: higher energy prices weigh disproportionately on people with low incomes. If we wish to encourage sustainable economies in fossil energy, the use of taxes or carbon pricing is more effective.

A great many fiscal systems include measures that support the use of fossil fuels by capping their selling prices or subsidizing their production. The amounts concerned can be considerable, as for example in such countries as Iran, Indonesia and Ukraine. But many other countries practise this type of perverse incentive in a smaller way, including Germany, which still finances its coal mines. The elimination of this type of subsidy is a priority for improving energy efficiency.

The tax system can also be used in the right direction. The introduction of subsidies or tax credits for investing in energy-saving equipment is one lever. The use of this instrument may be justified for encouraging the development of new energy-saving equipment. In practice, it can be costly for public finances and somewhat unfair in that it favours better-off households. The use of taxation to raise the price of using energy from fossil fuels is more effective and fairer if the revenues generated are judiciously redistributed. Most effective is basing this type of taxation on the carbon content of the various fuels. This is the principle of the carbon tax that has been successfully introduced by certain northern European countries, notably Sweden, Norway, Denmark and Ireland.

Differential pricing of products taking account of their use costs would be a more powerful lever, though unfortunately one that has been very little explored. If the cost of energy use were to be incorporated into the selling price of goods, this would unquestionably alter purchasing habits. In the vast majority of cases, the cheapest washing machines or kitchen stoves are the least energy efficient. The saving made at the moment of purchase is quickly eaten into by the additional cost of electricity consumed when they are used. Once again, it is largely low-income households that find it difficult to afford expensive equipment which are penalized.

Another important aspect of pricing concerns electricity. California, for instance, has set up a progressive pricing system that gives incentives to power companies to encourage their customers to save electricity. The system of 'energy performance certificates' operating in the UK, and less successfully in France, is intended to produce the same type of incentive. In the future, electricity pricing is likely to become much more sophisticated with the spread of 'smart' meters offering more flexible grid management.

Price incentives are just one part of the spectrum of instruments that can be deployed to economize on energy. They are most effective when combined with suitably conceived standards: building codes, consumption and emission standards for vehicles, technical standards in professional equipment and household appliances, and so on. Consider, for example, the consumption standards that are mandatory for light vehicles. Until 2008, such standards were, for example, much less strict in the US than in Europe or Japan. The result was incontrovertible: in 2006, the fleet of new vehicles put on the market in Japan and Europe emitted on average 150 grams of CO_2 per kilometre covered compared to more than 210 grams in the US.

In the longer term, urban regulations also play a decisive role. The fragmented city, with sprawling housing developments extending indefinitely, is doomed to energy overconsumption. Take Atlanta and Barcelona, two cities with similar populations. Yet Atlanta is twenty-four times larger in geographical area than Barcelona and emits seven times as much greenhouse gas. This situation was not freely chosen by its inhabitants, but is a constraint imposed by spatial planning. Thus we see here a powerful limitation to 'citizen voluntarism' in economizing on energy use.

The contribution of renewable energy

Although it is essential, investment in energy efficiency is very unlikely to be sufficient. It must be supplemented by substituting energy sources that do not emit CO_2 for fossil fuels. The portfolio of known technologies or those in development is greater for electricity generation and the production of heat than for fuel for transport, which remains predominantly of fossil origin.

Hydroelectricity is the most commonly used renewable energy. Norway, despite its large gas resources, uses it to produce more than 90 per cent of its electricity supply. Hydroelectricity is easy to secure a return on if investors are patient: a dam shares the same economic model as a toll-paying motorway or a car park. The fixed costs are high, but once the investment is amortized, it becomes a highly efficient machine for generating cash flow. Large dams give rise to many problems, and are the target of recurrent objections on the part of Western ecologist associations. Such opposition has not prevented the Chinese, who have no need of Western backers, from constructing the Three Gorges dam. The attraction of the hydraulics practised with dams is to be able to store surplus electricity in order to meet peaks in demand, a major advantage over wind-generated and solar energy.[6]

Obtaining energy from wind has been practised for a long time on land: the current term is onshore wind power. This form of renewable energy is coming close to the profitability threshold in many cases in Europe with the price of CO_2 around €20 a tonne. The development of onshore wind power comes up against the problem of intermittence. Wind power is very useful as a decentralized source or as back-up energy. But if one wants to use it for baseload production, one is obliged to double capacity to maintain the supply when the wind drops or becomes too strong.[7] Despite requiring additional investment and being economically onerous and ecologically uncertain, Denmark produces 15 per cent of its electricity from wind power. As soon as the wind drops, it imports large amounts of power generated by German coal-fired thermal installations. Offshore wind power, which is considerably more expensive, is still in its infancy and its potential is difficult to evaluate. Although wind conditions are generally better at sea than on land, a lot will depend on how the technology

---

[6] Some hydroelectricity is obtained from 'run-of-river' techniques, where the energy produced by the turbines is dependent on the river current. With a dam, the operator can quickly increase or decrease production by releasing more or less water. If the installation is equipped with pumps, the water level in the reservoir can be raised during non-peak hours when the grid has excess kilowatt hours (KWh). This is one of the few ways of storing electric power on a large scale.

[7] It is not a matter of doubling in a strict sense. In France, simulations carried out by Réseau de Transport de l'Electricité (the French transmission system operator) engineers shows that a wind power generating capacity of 10,000 megawatts (MW), judiciously placed across the territory, brings a guaranteed equivalent to 2,800 MW to the grid.

develops. One possibility for the future might be floating wind farms that could simultaneously recover energy from marine currents.

Another source of renewable energy, available in practically unlimited quantities on the human scale, is solar power. The energy in the sun's rays falling on the Earth's surface every hour is equivalent to the world's total energy consumption over a year. Mankind currently exploits only an insignificant fraction of this energy. And it is not for want of trying: the first solar power generator was built in 1912 in Maadi to the south of Cairo. Designed by the American engineer Frank Shuman, it was supposed to drive a 55-horsepower steam pump for drawing water for irrigation. But Shuman soon went bankrupt. In the 1980s, Lutz Inc. built nine solar power generators in the Mojave desert, California. In 1991 this also went bankrupt after subsidies from the state of California were reduced.

Solar power will probably become one of the pillars of the energy system in years to come. But its deployment on a large scale is dependent on reducing costs that are still much too high. Until now, its use has grown where subsidies have been available: there are three times as many solar panels in Germany as in the whole of the Mediterranean region. Cost reductions are currently being made in two major technologies in the sector: thermal, which directly converts the sun's rays into heat, and photovoltaics, which transforms them into electricity through semiconductors. German followed by Chinese manufacturers have acquired a comfortable lead in first-generation photovoltaic technology. The stakes are much more open for the manufacturing of thin-film solar cells, which could play a major role in the spread of this technology, and where manufacturing methods are achieving much better efficiencies. There remains one big unknown: whether they can be produced on an industrial scale at a lower cost than crystalline silicon cells.

A feature of renewable energy sources is that electricity can be generated in a decentralized manner. But their development comes up against all kinds of organizational obstacles or sometimes more prosaically the hostility of established grid operators. For the electrification of remote rural areas they can bring substantial economies over investing in distribution grids. This advantage has only been marginally used to improve access to electricity for rural populations in the least developed countries and those with a low population density.

The portfolio of technologies for producing non-emitting fuels for transport is more limited than for electricity production. Hydrogen engines using a non-emitting fuel cell are at the prototype stage. Furthermore, a lot of energy is required to produce the hydrogen. Electric engines come up against the problem of battery autonomy, again because of the difficulty of storing electricity, and other more practical considerations. For example, it takes only a couple of minutes to fill up a car with petrol, but several hours are needed to recharge the batteries of an electric vehicle. The most tried and tested technology is biofuels. But their development, like bioenergy more generally, should be very carefully analyzed: biomass is not a renewable source independent of human activity, such as the sun's radiation, wind power, marine currents or the heat in the earth's crust used by geothermal technology.

Good use of biomass

Biomass comprises all organic matter of plant or animal origin. By extension, it refers to all the energy – referred to as 'bioenergy' – that can be recovered from different organic materials, such as wood, household waste, agricultural waste, etc. Bioenergy can come directly from the combustion of wood or other plant matter. It can also result from the recovery of biogas produced when organic waste decomposes. It can be used after plants have been converted into liquid fuels – in which case we speak of 'biofuels'. It could also at some future point be derived from seaweed, a source the potential of which has been little explored.

When biomass is obtained from plants that reproduce them-selves, for example from a forest where trees are replanted, it can be described as renewable. When this is not the case, and resources are removed without ensuring that they are replaced, it disrupts the car-bon cycle in a similar way to the use of fossil energy. Biomass then becomes a net emitter of CO_2, but for reasons of accounting conven-tion its emissions do not appear alongside those of the energy system. Instead they are included within emissions from the agriculture and forestry system.

In 2006, biomass was the fourth largest source of primary energy, after the three fossil fuels. With a use equivalent to 1.2 million tonnes of oil, it represents three times the total contribution of renewables. Forecasts by the IEA show that it could make a huge contribution to the reconfiguration of the energy system. In the most favourable scenarios, biomass supplies nearly a quarter of the primary energy used worldwide in 2050, way ahead of all other renewable energy sources. To achieve this, investment must be made both in the con-version of the traditional systems used on a very large scale in poor countries and in new uses in which bioenergy is progressively substi-tuted for fossil energy.

So-called 'traditional' biomass results from taking material directly from the natural environment by poor rural populations that do not have access to other forms of energy. In the least developed countries such as Nepal or some sub-Saharan African countries, its contri-bution approaches 80 per cent of total energy supplies. Because of demographic pressure and the degradation of ecosystems, a high pro-portion of traditional biomass can therefore no longer be described as renewable. The reconversion of these systems requires quickly find-ing new energy sources to reduce the pressure on the environment and to meet people's basic needs. In the short term, this will rely in part on switching to fossil fuels. But the impact on global emis-sions will be relatively small: the World Bank has calculated that connecting to the grid the 1.6 billion people in the world currently without electricity would emit the same amount of greenhouse gas as could be saved in the US by converting its light utility vehicle fleet to European standards.

The use of biomass as a substitute for fossil energy firstly concerns wood. In the current technical conditions, the use of wood to produce heat or electricity tends to become economically viable with a relatively modest carbon price. This is the case in the European electricity sector in which operators are partially switching from coal to wood in large thermal power stations. The growth of this type of use will require the creation of supply channels and of substantial logistical installations. The costs of preparing, transporting and storing biomass are in fact high and account for a very high proportion of its cost price. It should be added that agricultural residues are a potential source of biomass that is largely under-used.

The recovery of biogas from organic waste offers interesting possibilities that are easy to make cost-effective with a moderate carbon price. But the potential is not unlimited: the growth of organic waste is slower than that of packaging waste.

There remains the controversial issue of biofuels. The first industrial biofuel process was developed by Brazil after the second oil shock on the basis of using sugar cane, which has excellent energy yields. Largely subsidized at the outset, this industry becomes relatively profitable once the price of a barrel of oil reaches around $50. And it does not directly eat into tropical forest. Biofuels have recently become heavily subsidized in the US and, to a lesser extent, in Europe, where their use has been strongly encouraged. In the US and Europe, biofuel technology seems, however, to be much more economically and ecologically costly than the use of biomass for heat. Rather than try at any cost to run cars on maize or colza, it would make sense to improve the exploitation of forests so as to obtain more wood to use as a fuel in boilers. In addition, the effect of biofuel production on greenhouse gas emissions is uncertain. But the main barrier to expanding it further lies in the pressure it puts on food production. The growth of biofuels contributed to the tension on world agricultural markets in 2006–8 and to the food crisis that resulted from it.

The margin for development of first-generation biofuels is therefore limited. Second-generation technologies are set to convert entire plants into fuel, rather than conventional crops as now. They could offer a greater potential for development. In addition, there are many species of seaweed that can produce oil that is usable as fuel. In laboratory trials some species have been shown to be capable of producing up to 800 times as much oil per hectare as soybean or colza.

The contribution of nuclear energy

Another controversial issue is nuclear energy. Following the Three Mile Island accident and Chernobyl disaster, its deployment was stopped in some countries and reduced in others. More than 60 per cent of the total number of nuclear installations are found in the US, France and Japan. But most extensions of nuclear capacity today are occurring in Asia and eastern Europe. Nuclear energy remains a high priority for public research and development programmes, way ahead of renewables and energy efficiency. Around $10 billion will, for example, be spent at Cadarache, southern France, in the framework of the International Thermonuclear Experimental Reactor (ITER) international programme for testing the feasibility of electric power production from nuclear fusion.

Nuclear energy is a very capital intensive technology. Part of its cost is a provision for the future – decommissioning end-of-life power stations and radioactive waste management – which is hard to estimate and is often concealed to a greater or lesser extent by operators. It is a very inflexible technology, exclusively suited to the generation of baseload power. It is also a technology which has risks associated with its deployment: the risk of accidents in power stations; risks linked to the storage of the radioactive substances present in nuclear waste; and the risk of the spread of fissile material that can be used to make explosive devices. The first two types of risk can be better managed with the aid of technology. The third is a threat against which it will be more difficult to provide protection. Consequently these various risks will need to be carefully weighed up against the benefits of nuclear energy, i.e. the production of baseload electricity without carbon emissions and reasonably inexpensive.

When coal cleans its waste

Let us take the most favourable scenario, that all the world's countries forget their divergent interests and manage to introduce a carbon price that applies to all actors worldwide producing energy from fossil fuels. Let us say this is agreed at €100 a tonne in 2030 and €300 a tonne in 2050. At the same time they initiate ambitious standards and research and development (R&D) policies to reduce emissions. Would this enable all fossil-sourced energy to be taken out of the world energy mix by the end of the century without threatening the energy security? According to the experts of the IEA, the answer is no. Even in the most optimistic scenarios, it is doubtful whether the world's energy needs can be met without recourse to coal, oil and gas. Thus we have to continue investing in fossil energy, but in a different way.

Let us first look at coal. Proven coal reserves amount to more than 170 years of production. They are better distributed around the world than oil reserves and are plentiful in three large countries that are key to climate negotiations, namely the US, China and India. The abundance of coal is reflected in very low extraction costs. The real cost of coal lies in its transport, storage and handling. It is naïve to imagine that all this coal will remain in the ground.

The use of coal will therefore continue for a long time. To avoid runaway climate risk, its use must rapidly be subject to a simple rule: the capture and underground storage of CO_2 emissions. In a few years time, such a rule will seem as natural to us as banning the discharge of waste water onto the public highway. Yet such a practice was standard in towns before the municipal authorities agreed to make massive investments in sewerage systems and waste water treatment. Carbon users should agree to the same type of investment to capture and then store CO_2 emitted from power plant chimneys.

The first carbon capture and storage experiment was launched at the Sleipner gas platform in the North Sea. Gas from the Sleipner deposit contains 9 to 12 per cent CO_2, which prevents its commercial exploitation as it is. Since 1996, the operator Statoil has been capturing a high proportion of it on the platform and then reinjecting it into an aquifer under the sea bed. The company processes a million tonnes of CO_2 a year, thereby preventing it from escaping into the atmosphere. The launch of the operation was facilitated through the existence in Norway of a €40 tax levied on every tonne of CO_2 emitted from oil and gas installations.[8] The reinjection of the unwanted gas thus brings savings of €40 million a year in taxes. Two other installations producing natural gas have introduced similar systems, in Salah, Algeria and Snohvit in the Barents Sea. Many other pilot projects are under development and it is difficult to forecast how quickly this procedure will become standard practice in the industry.

The stakes are highest in regard to carbon capture and storage for electricity production, which already accounts for 70 per cent of the carbon produced worldwide. The technology for achieving this breaks down into three separate stages: the capture of CO_2, transporting it, and then storing it underground. There are three competing technologies for capturing CO_2. All of them allow nearly 100 per cent of the carbon emitted to be captured, which is important in environmental terms. But because they reduce the energy yield of power stations they are costly to implement. They therefore

---

[8] The system was altered in 2008, with the entry of Norway into the European Union emissions trading scheme.

require the use of more coal to produce the same amount of energy. CO_2 capture accounts for 70 per cent to 80 per cent of the total cost of the investment. Transporting CO_2 presents no particular problem and the use of gas pipelines is likely to be the preferred method. Finally, underground storage reservoirs able to resist the passage of time need to be found. According to the geologists there is no shortage of these. All that remains then is to agree on monitoring methods that eliminate the risk of any future release of the CO_2.[9] This also raises questions of the sharing of responsibilities: who will be responsible in 500 years' time for the CO_2 that we pump underground today?

After the ten or so years still needed to set up the industrial processes involved, the rate at which carbon capture and storage spreads will depend largely on economic conditions. In Europe, where electricity generation already comes within a CO_2 cap-and-trade system, a carbon allowance price of €60 is currently estimated as the level above which production with CO_2 capture and storage becomes more cost-effective in new power stations. As time passes, this price will sharply fall for new installations, but it will always be much more expensive to equip old power stations.

Once they become commonplace, carbon capture and storage technologies can also be used by gas-fired power stations and even those operating with biomass. If this biomass is renewed, energy is then produced while storing atmospheric carbon in the ground.

A fairly moderate carbon price will make the general use of carbon capture and storage economically viable in power generation. The

---

[9] There are two kinds of risk. One is the possible ineffectiveness in regard to controlling emissions if leaks allow CO_2 to escape imperceptibly into the atmosphere. The other, in the event of large-scale releases, is the health risk that can arise for local populations. Accidents involving CO_2 are very rare, in contrast to the frequent occurrence of accidents that result in carbon monoxide (CO) inhalation. The most dramatic such accident occurred in 1986 on the shores of Lake Nyos, in Cameroon, where a sudden emission of CO_2 released from the lake led to the death of 1,700 people.

general adoption of carbon pricing in the coal sector leads therefore to two kinds of substitution: the first, which is well-known, from coal to energy that emits little or no carbon; the second from 'dirty' coal to 'clean' coal, with the unwanted CO_2 being returned where it came from – beneath the ground.

Carbon prices and oil rents

The future of oil appears in a different light. As we have seen, the main energy prospect for oil is the transport sector. Capturing and storing CO_2 emitted from cars and planes is completely unrealistic with known technologies. Of course, this situation may change in the future. But one cannot build long-term climate strategies on such conjectures.

The perspectives for the development of oil production depend on very different constraints from those applying to coal. Proven oil reserves are estimated at around forty years' extraction at the current rate. Three-quarters of these reserves are located in the Middle East. Since the oil era began, we have extracted half the total volume of oil available in known oilfields, a state of affairs that corresponds to the well-known 'peak oil' after which yields from wells start to decline.[10] Does this mean that the planet will run out of oil? Probably not. There are still little-explored oil deposits that could be opened up and exploited, even if access is more difficult and sometimes dangerous for the environment. Think of the oil lying beneath the Amazonian rain forest or the Arctic ice. There

---

[10] The notion of peak oil was developed by the geophysicist King Hubbert, who tried to generalize an observation that he had made in the American oil industry: when half the deposit of an oil well has been extracted, a peak is reached after which the technical exploitation conditions slow down exploitation and lower production. On the basis of these observations, in 1956 Hubbert drew a curve predicting that oil production in the US would peak in 1970, which is close to what actually happened.

is also a vast amount of so-called unconventional oil present in bituminous sands and oil shale. If all these resources are included, oil reserves increase fourfold, to give 170 years of extraction at the current rate.

Similarly, the increase in the price of oil facilitates the extraction of unconventional gas by high-pressure hydraulic fracturing of rocks. This operation, which is costly in terms of the environment, is already practised on a large scale in the US. Its widespread use could lead to a large upward revision of the amount of potentially usable gas in the world.

In a world without carbon constraint, the increased scarcity of oil through the progressive depletion of existing deposits leads to rising prices per barrel, which makes the exploitation of unconventional resources economically viable. The process continues as long as oil is more cost-effective to use than substitutes for it. It stops once it becomes cheaper to run cars and planes on biofuel or hydrogen rather than on oil extracted from under the icecap or from Canadian or Venezuelan oil sands. But as the alternatives are limited in the transport sector, the future looks rosy for oil prices and producers sitting on large reserves can expect to go on getting high revenues from the oil remaining in their wells.

The introduction of a price for carbon disrupts the oil rent system. If setting a price for carbon becomes standard practice, it leads to higher prices for the fuel used for transport without any increase in the price paid to the producer. Raising the price of carbon eventually reduces demand for oil products. This weakening of end-use demand has repercussions on the crude oil market, where the price falls (unless maintained by a supply-side cartel). Cheaper oil reduces the incentive to exploit new deposits. All in all, a rise in the carbon price lowers the price of crude oil, which acts as a disincentive for producers to develop and exploit the most expensive – and also the dirtiest – oil sources. While ecologically virtuous, this runs totally counter to the interests of producers.

Accelerating the energy transition

Made possible by the use of coal and subsequently oil and gas, the industrial revolution led to a sea-change in the energy system. This transformation resulted in the massive growth in sourcing by the rich countries, whose present billion inhabitants are dependent on unwieldy and relatively inflexible infrastructures that are heavily dependent on fossil fuels for their energy supply. At the other extreme, most of the billion inhabitants of the least developed countries lack access to minimal energy requirements. Between these two extremes, the emerging world numbers more than four billion people who aspire to join the comfortable lifestyle of developed countries by wagering on accelerated growth and an expansion of their energy supply.

The economic take-off of the emerging countries is a shock that is speeding up the long-term transformation of the energy system. It exerts new pressure on fossil energy stocks, which in turn aggravates geopolitical and economic tensions. The resulting upward trend in energy prices can cause serious economic damage, such as the great recession of 2008–9, which was partly triggered by the soaring price of oil. It also deprives two billion of the world's poorest people from access to energy. In the long term, rising fossil fuel prices encourage the opening up and exploitation of ever more numerous deposits and slow the much-needed transition to an energy system based on diversified sources.

On the contrary, taking climate risk into account could speed up this transition. Letting the world energy system react solely to energy prices results is an unconscionable risk from the standpoint of the climate: 6 tonnes of CO_2 emissions from energy use per capita in 2050, when it is essential not to go above 2 tonnes.

To limit it to two tonnes, energy investment in the rich countries must be reoriented without delay. To hasten the spread of technological innovations that economize on emissions, it would be advisable simultaneously to expand public research. Organizational innovations, particularly in relation to the functioning of cities and of transport and logistical networks, are also of strategic importance. Innovation also needs to be focused on developing new low-carbon energy systems. If these are economically attractive, they will become the future for emerging countries and will widely be adopted. The amount of CO_2 entering the atmosphere between now and 2050 depends primarily on the choices made by these countries. A strategy of breaking with the existing system in favour of a low-carbon energy system does not at all create a dilemma in regard to the aims of eradicating the great energy poverty that currently exists in the world. On the contrary, it offers new options.

Such a drastic reorientation of our energy choices cannot be made within the existing framework of economic rules, where prices reflect the greater or lesser scarcity of fossil energy stocks in the ground. Such prices do not take into account another scarcity, namely that of the atmosphere, which will not be able to absorb all the carbon currently trapped within these stocks. From the economic standpoint, this new scarcity of use of the atmosphere has a price: the price of carbon which must become the compass guiding tomorrow's choices.

FOUR

Pricing carbon:
the economics of cap-and-trade

Lodz Province, Poland. The town of Kleszczow has a little over 3,000 inhabitants, most of them farmers and retired people. It is part of the district of Belchatow, in the historical industry centre of Poland. Kleszczow is home to the largest coal-fired power station in Europe, its two main chimneys each as high as the Eiffel Tower creating a land-mark that can be seen from a great distance on clear days. Every year Belchatow power station generates a fifth of Poland's electricity from its twelve generating units. It is owned by Polska Grupa Energetyczna (PGE), a company which also owns two lignite mines – lignite being the most carbon-emitting coal. This vertical integration ensures the plant's supplies. Belchatow also emits more carbon dioxide (CO_2) than any other installation in Europe, around 30 million tonnes a year, or a third of the total emission of countries such as Portugal or Colombia.

Yorkshire, UK. Drax lies six miles south of Selby, famed for its Benedictine Abbey dating from 1069. The village had only 382 inhabitants at the last census. Since the closing of the village shop in 2007, which also served as the local sub-post office, the Huntsmans Arms pub is the only remaining business. Yet Drax is the site of the largest coal-fired power station in the UK. Its six generating units, with a total capacity of 4,000 megawatts (MW), supply 7 per cent of the country's electricity. The power station can process 36,000 tonnes of coal a day. Following major investment in desulphurization equipment, in 1998 Drax became the 'cleanest' power station in the United Kingdom. Yet though local pollutants have been reduced to a minimum, CO_2 emissions have been unaffected. Every year Drax discharges more than 20 million tonnes of CO_2 into the atmosphere, equivalent to a quarter of the total emissions from the country's cars.

Since 1 January 2005, Drax and Belchatow power stations have both come within the European Union emissions trading scheme (EU ETS). Over the first five years the system has been in operation, Drax has been the leading buyer of emission allowances, while Belchatow has been alternately a buyer and seller on this new carbon market. Like 11,000 other industrial installations in Europe, they have left the common economic regime in which emitting greenhouse gases into the atmosphere free of charge was an unlimited right. For Drax and Belchatow, carbon now has a price.

How to price CO_2

There are two possible ways of pricing CO_2: taxation or an emissions trading scheme.

The use of a tax for protecting environmental resources was advocated as early as the 1920s by the English economist Arthur Pigou. His idea was to protect the environment by incorporating into the costs of production an estimation of the value given to the protection of environmental assets by society. At what level should this tax be set? In purely theoretical terms, the answer is straightforward: the level at which the marginal cost of the tax equals the environmental advantage aimed at. The more value society attaches to the environment, the more it will be willing to agree to a high tax, and vice versa. In practice, economists have considerable difficulty in answering this question in regard to CO_2 emissions. When one wants to introduce a carbon tax, one usually just estimates beforehand a tax level enabling emissions to be reduced by a certain amount.[1]

The advantage of a carbon tax lies in the visibility of the price of carbon, which is known by everyone. Its disadvantage is that it is difficult to anticipate the effects of this new price on the volume of emissions. The environmental result of this type of policy is therefore uncertain, at least at the moment when it is implemented.

Emissions trading was explored in the 1960s by economists Tom Crocker and John Dales, following the pioneering work by the Nobel Prize-winning economist Ronald Coase. The move from theory to practice really began in 1995 with the US, in the context of the fight against acid rain, when a market in the right to emit sulphur dioxide (SO_2) into atmosphere was successfully launched.[2] Emissions trading schemes are a recent innovation, which explains why their function-ing often remains rather poorly understood.

The basic mechanism is not rocket science. In emissions trading, the volume of authorized emissions is set in advance, then one leaves it up to the market to determine the price. The expression 'cap-and-trade' sums it up concisely. An emissions trading scheme implies first of all that scarcity is created by capping the authorized emissions. Once this cap has been set, one can then trade the rights to emit, which have become scarce due to the new carbon constraint placed on economic players. What is the advantage of a cap-and-trade sys-tem? Quite simply to implement emissions reductions where they are

---

[1] In terms of techniques, Pigou recommends using a 'cost–benefit' method, but in practice a 'cost-effectiveness' method is used, which, if the marginal abatement cost curves of reduction are estimated correctly, allows a given emissions reduc-tion target to be achieved optimally in economic terms.

[2] Acid rain is one of the main consequences of atmospheric emissions of SO_2. In the context of the Clean Air Act, the US put a cap on emissions from power stations, which are the main source of SO_2, and set up an emissions trading scheme oper-ating at the federal level. This market more than halved the cost of the emissions reductions obtained.

least expensive. The economic efficiency of action against climate change is thereby increased. The environmental ambition of the system results from the *ex ante* setting of the total emissions level, the cap. The role of emissions trading is then to ensure the economic efficiency of the methods used to attain this objective. We are in a symmetrical situation to that of a carbon tax, with knowledge of the outcome of the environmental policy adopted, but uncertainty as to its cost for the economic actors. Must we then favour one or another of these instruments?

Tax or emissions trading?

In the public debate around this question, there are first of all a number of easily made but flawed criticisms at an economic level. Let us begin by dealing with these.

The argument is often put forward that introducing a carbon tax involves imposing a new charge that is liable to weigh down the economy. In point of fact, the macroeconomic impact of a carbon tax depends entirely on the circumstances under which it is introduced. For example, Sweden introduced a carbon tax on consumers of fossil-derived energy in order to reduce other taxes imposed on factors of production such as labour or capital.[3] Such an impact can, on the contrary, boost the economy. Economists refer to this as a 'double dividend', in which there is not only a benefit to the environment but also a positive economic impact.

Emissions trading schemes are sometimes viewed as a perverse mechanism that 'gives' new rights to pollute to the actors concerned. In fact, Drax and Belchatow power stations had no emission allowances

---

[3] Sweden experienced a very serious economic and financial crisis in the early 1990s. In response to this, it completely changed its tax system by reducing the overall level of its tax and social security deductions, in particular by the reduction of social security contributions and of corporation tax. Simultaneously it set up an environmental tax system, with a carbon tax as one of its key elements.

prior to January 2005. Did the launch of the European market give them new rights to emit CO_2? Quite the reverse: previously they had unlimited rights and the introduction of the market restrained them by setting a cap on emissions. Put simply, the unlimited right to emit CO_2 is an implicit right. Setting up a cap-and-trade scheme restricts this right by instituting an emissions cap that makes explicit rights appear.

Can economists help decide between the respective merits and drawbacks of the tax and the market? At a theoretical level, there is general agreement that the two instruments would be strictly equivalent in a world where markets functioned perfectly and where there was no uncertainty: the public authorities would set the tax at exactly the price level that appeared on the market given the agreed-upon cap. It therefore makes no difference whether they set the price (the case of the tax) or the quantity (the case of the cap-and-trade).

There is no need to remind anyone of the degree of uncertainty decision-makers are surrounded by in regard to climate change. This uncertainty complicates economic reasoning and makes its conclusions more controversial. Martin Weiztman (1974), a professor at Harvard, has provided us with the standard theoretical analytic model. In regard to climate change, Nordhaus deduces from it that a tax would be preferable to an emissions trading scheme. Stern challenges this conclusion.

The truth is that theoretical considerations have hardly any influence on the choices made in regard to carbon pricing on the ground. First of all, the two instruments are not necessarily in competition with each other. In fact, a tax and a cap-and-trade system can, as is currently the case in most northern European countries, be combined. Secondly, the system adopted is the one that succeeds in foiling conservative opinion and established interests by forming the right coalition of actors. And here, political sociology provides more keys to understanding than economics.

To set up a carbon price, Europe successively tried two methods. In 1992, the European Commission filed a harmonized tax project on industrial CO_2 emissions, hoping to be able to use the provisions of the single market for taking qualified majority decisions on environment matters. This project came up against considerable hostility from industrial lobbies and a majority of EU countries, which were against the idea of giving up part of their innate right to levy taxes. Confronted by the massed ranks of the opposition, the planned carbon tax project was first passed over to discussion by experts in order to harmonize the indirect tax thresholds levied on fuels. These discussions soon became bogged down, and the Commission formally withdrew the proposed directive in early 1998. In the meantime, Europe had signed the Kyoto Protocol in December 1997, some of the measures of which anticipated using market mechanisms. Forced to change direction, Europe then decided to follow the emissions trading road to achieve its objectives of reducing greenhouse gas emissions.

The overall architecture of the market

The EU's climate policy is officially committed to the objective of helping limit the average warming of the planet to 2°C. But this does not include a long-term commitment in terms of capping global emissions. On the other hand, Europe has made such commitments in the medium term. When it signed the Kyoto Protocol, the EU committed itself to reducing its average greenhouse gas emissions over the period 2008–12 by 8 per cent compared to 1990 levels. In December 2008, it committed itself to reducing its emissions in 2020 by 20 per cent compared to 1990, with the adoption of what is generally known as the 'climate and energy package'. This commitment is unconditional and is registered with the United Nations (UN). It can be raised to 30 per cent in the event of a 'satisfactory' international climate agreement. The carbon market is the joint economic instrument set up to achieve these objectives at the least cost.

The EU ETS applies to a little over 11,000 industrial installations in Europe and covers their CO_2 emissions.[4] These installations account for about half the total European CO_2 emissions. With more than half of the allowances allocated, the power sector is by far the main sector concerned. Other energy producers, oil refineries and heat generating plants account for a quarter of the allowances allocated. The remainder are basic industries that use large amounts of fossil fuels: steel production, cement, glass and ceramics, paper and cardboard manufacturing, and all other combustion plants with more than 20 MW of installed power. It was decided to include large industrial installations only, since their emissions are much easier to calculate and check than those coming from the many small-scale emitters. Concretely, this has led to the exclusion of emissions from the transport, agriculture, construction and waste management sectors.

The trial period covered three years, from 2005 to 2007. The permitted emissions cap was not defined in advance in a centralized manner, but resulted from negotiations between the Commission, the member states and industry, and overall was set at a moderate level. During this period, most allowances were allocated freely, which greatly facilitated the launch of the market.

The second phase, from 2008 to 2012, overlaps with the implementation phase of the Kyoto Protocol. The constraint on industry was raised: overall, the authorized annual emissions cap was lowered by a little less than 10 per cent compared to the previous phase, with the situation varying widely from one country to another. One important characteristic of this period is that, to meet compliance, firms may use carbon offsets, which are investments in emissions reduction projects implemented under the Kyoto Protocol. Limited to 13 per cent at the

---

[4] The system was launched in 1995 with the twenty-five member states of the EU. Bulgaria and Romania joined it in 2007, and Norway, Liechtenstein and Iceland in 2008.

EU level, this mechanism has greatly facilitated the launch of the international carbon offsets market.

The third phase has been extended to eight years running from 2013 to 2020. With the inclusion of air transport and of greenhouse gas emissions other than CO_2 for the aluminium and chemicals industries, the market coverage will be enlarged by a little more than 10 per cent. The emissions cap, set on a centralized basis for the whole of Europe, is expected to fall by a little less than 2 per cent a year from 2013 onward. The constraint weighing on European industry is thus well known in advance. Another major change concerns the allocation of allowances. During the first two stages, free allocation of allowances was very much the norm. During the third stage, auctioning is expected to become the general rule – from 2013 in the electricity generating sector, and more gradually in the industrial sectors where the shift to auctions is liable to disrupt the conditions of competition outside the EU.

At a technical level, the credibility of the constraint imposed on producers is based on a fairly sophisticated mechanism, the central feature of which is the register system whereby all emissions permits traded between market participants are recorded. These national registries are a kind of large-scale accounting record in which each installation must open an account in which its CO_2 allowances are issued. Each allowance is a permit that gives the right to emit one tonne of CO_2. The sum total of allowances allocated constitutes the authorized emissions cap. Every allowance is carefully listed and can be tracked through its identification number. A purchase of an allowance is recorded as a credit in one's account and a sale as a debit. Once a year, installations have to surrender the number of allowances corresponding to the CO_2 they have emitted. The returned allowances are then permanently erased from the registries so that they cannot be used a second time. This operation to ensure compliance allows the total amount of emissions produced and the volume of allowances refunded to be known. This information is required to be publicly made available. It is invaluable for the proper working of the market.

As with any reliable accounting system, outside auditors, independent of the companies concerned, need to check the accounts to ensure that the information provided is correct. An auditor accreditation system has been established. In addition, there is a penalty for non-compliance: since 2008, a fine of €100 a tonne is imposed on any operator who has failed to pay back the number of allowances corresponding to his CO_2 emissions. One does not treat the law lightly! We shall now see how this system works in practice.

Looking for 22 million tonnes in CO_2 allowances

Like the 11,000 other installations covered by the system, every year, in February, Drax power station receives its annual CO_2 allowances. These allowances, allocated free of charge, represent its emissions cap and are recorded in its account in the UK emissions allowances registry. The cap is determined according to the method based on its previous emissions, the so-called grandfathering method: during the trial period, the UK administration calculated the right to emit by power plants such as Drax by taking as a reference point the emissions from the period 1998–2003, minus the level of effort demanded from the UK electricity sector as a whole.

Every year, before the end of April, the power station must surrender the number of allowances corresponding to its CO_2 emissions the previous year. If it emits less than its cap, it can sell its surplus allowances on the market. Or, it can keep its unused allowances for the following year. In this case, it saves up allowances in the expectation of future obligations or because it anticipates an increase in the price of carbon. Such saving is most commonly referred to as 'banking'. If Drax emits more than its cap, it has to buy allowances from another installation. It can also use some of the allowances it received in February, though such 'borrowing' will make compliance more difficult the following year. As with banking, such borrowing of allowances is not permitted between the first and second periods.

From 2005 to 2007, Drax emitted slightly more than 65 million tonnes of CO_2. The power station's activity was sustained by a strong demand for electricity. With its present installations, it has no alternative to coal for producing its electricity. Since its emissions cap was set at just over 43 million tonnes of CO_2, Drax had to find 22 million tonnes worth of allowances in the market to be in compliance with the EU ETS rules. Over this period, it was the leading purchaser of carbon permits in the European market. The company had trouble obtaining allowances from other UK installations because the country as a whole had an allowance deficit. Most of the allowances were found from installations in other countries. Poland, along with Estonia and France, was one of its main suppliers.

Disposing of excess allowances

Belchatow power station received an allocation of 92 million tonnes of CO_2 for the whole of 2005–7. In 2005, Belchatow's emissions were slightly higher than the allowances it had been given. Then in 2006 and 2007 its emissions fell thanks to bringing on-stream equipment that enabled the plant's efficiency to be improved by 3 per cent. Its emissions went below its cap, and over the three-year period Belchatow was, through its surplus allowances, able to put 2 million tonnes of CO_2 onto the market. With the tightening of the constraint in the second period, Belchatow once again fell short of allowances in 2008.

Other Polish power stations have profited to a greater extent from the carbon market. For example, the Polaniec power station, owned by Electrabel since 2003, made investments enabling it to add up to 20 per cent wood to its fuel. This wood–coal mix does not reduce the energy performance of power stations so long as the proportion of wood does not exceed 20 per cent. It increases costs, but this increase is covered by the price of carbon once this reaches or exceeds around

€20 a tonne. Burning wood, and more generally any biomass fuel, is deemed not to be CO_2-emitting under the rules of the EU ETS: the CO_2 emitted is balanced by the absorption of carbon by trees and plants as they grow. By systematically using wood, Polaniec was able to put on the market more than 6 million of the 21.75 million allowance it had been allocated over 2005–7.

Power stations such as Drax, Belchatow and Polaniec are the heart of the European carbon market. Overall, the electricity generating sector has been a net buyer of allowances. The market is said to be 'short' in allowances. Other industrial sectors have as a whole received more allowances than they have emitted CO_2. They have been 'long'.

As we have seen, the member states of the EU had great independence in the initial allocation of allowances. They generally were careful to avoid any risk of their local industries becoming less competitive. Heavy industry sectors, which are highly concentrated and subject to fierce international competition, were thus treated generously. Take the example of steel production. From 2005 to 2007, the industry posted record levels of production in Europe. It therefore had a surplus of allowances. The Arcelor-Mittal blast furnace at the port of Dunkirk is the top CO_2-emitting installation in France. It had 1.6 million tonnes of surplus allowances during the trial period. The Arcelor-Mittal group as a whole received 240 million tonnes of allowances for its actual 180 million tonnes of emissions. It was a net seller of 60 million tonnes of allowances.

Many small and medium-sized installations also had surplus allowances. This is the case, for example, for most of the 210 urban heating installations in France, for which the total surplus was 7 million allowance units. In some instances, this surplus has been increased by taking action to reduce carbon emissions in response to the introduction of a carbon price. But in many other instances, the surplus allowances reflected the generosity of the original allocations.

The functioning of the carbon market can now be seen more clearly. Over the first three years, a cap of 6.23 billion tonnes of CO_2

was imposed on producers subject to allowances, which in fact emitted 155 million tonnes less than their overall cap. The market as a whole was in surplus. But at a microeconomic level, very few installations strictly matched their actual emissions to their allocations. A third of installations were short of allowances to cover their emissions, and their deficit reached around 650 million tonnes of CO_2. Two-thirds had more allowances than they needed, with a surplus of 805 million tonnes. The role of the market was to facilitate as efficiently as possible the transfer of allowances from surplus to deficit installations.

Buying and selling CO_2 allowances

How does one buy 22 million CO_2 allowances or sell 6 million? European regulations allow any EU citizen to open an account in a national registry and buy and sell carbon. Very few ordinary citizens are, however, familiar with this type of transaction. But for the professionals, the European carbon market has become a key market, on which more than €60 billion were traded in 2008 and 2009, or more than 80 per cent of the world's carbon trades.

To find 22 million CO_2 allowances, Drax power station can first try to obtain them directly from its counterparts. It then needs to contact other installations to suggest buying allowances from them. Such bilateral transactions represent less than 10 per cent of the market.

A second alternative is to turn to a broker, whose job it is to put buyers and sellers in touch with each other. In contrast to traders, brokers do not take positions on their own account. They act on behalf of their clients and are paid by means of a margin taken on the difference between the bid price and the asking price. In this way the market is kept fluid. Most trading of CO_2 allowances takes place through the intermediary of brokers. They are particularly partial to big-league players like Drax that have to trade large volumes, but are less interested in 'small accounts', a situation that explains the difficulty some installations have in gaining access to the market.

Bilateral and brokered transactions are both based on the mutual trust of the parties concerned. They are carried out 'over the counter', as opposed to transactions in organized markets.

The organized markets, or carbon exchanges, provide their participants with two types of services: continuous transparency of the listing of bid and asking prices, and the permanent possibility of buying or selling any quantity at these prices with a guarantee of allowance delivery and payment via a clearinghouse or its equivalent. Six carbon exchanges were launched during the EU ETS trial period. The two largest are the European Carbon Exchange (ECX), based in London, which dominates the forward contracts market, and Bluenext, based in Paris, which dominates the spot market transactions. Carbon, like any other commodity, can be traded on a spot or futures basis.

As from 2005, the managers at Drax could see that they were going to be short of CO_2 allowances over the 2005–7 period and even beyond that. A wait-and-see management style would have involved waiting for the compliance periods to find out the amount of missing allowances and then buying them for cash. This type of management gives rise to great uncertainty as to the final cost (or gain) of compliance. The derivatives market enables market participants to hedge in advance the risks of non-compliance and of fluctuations in permit prices. It has grown more rapidly than the spot market, accounting for three-quarters of carbon traded during the first phase of the EU ETS.

We are still a long way away from the degree of financialization of other commodity markets. In the oil market, for example, thirty-five forward contracts are traded for one barrel of crude oil. A proportion of these contracts no longer serves as hedging instruments for oil operators, but have become tools that financial players use to conduct purely financial operations. Should we be worrying about a possible 'carbon subprime' crisis?

Speculation, misappropriation, regulation

Like the oil market, the carbon market is a kind of hybrid. Spot trades concern allowances which are intangible products – the right to emit a tonne of CO_2 – marked by an entry in a national registry. This product is not a financial product, any more than a barrel of oil in a tanker. The contracts or options traded on the futures market are, on the other hand, financial products. They are therefore subject to the standard rules applying to financial markets, rules that have been fairly thoroughly re-examined following the damage caused by the 2008–9 economic and financial crisis.

When one speaks of the risks of dysfunction on the carbon market, what immediately comes to mind is speculation on derivative products carried out by greedy and unscrupulous financial players. In the case of the climate there is an added ethical component: we are hardly going to let traders make a killing speculating on climate risk faced by our children and grandchildren. During the first five years, serious dysfunctions did, however, arise on the spot market, which is not subject to any specific regulation.

Firstly, there was the value-added tax (VAT) fraud that developed on a large scale in 2008 and continued until governments pushed through emergency changes to the rules in the spring of 2009.[5] How much did this involve? It probably amounted to several billion euros, stolen from the tax authorities by criminals who had entered the spot carbon market by opening accounts in national registries. This kind of VAT fraud is difficult to detect for goods such as CO_2 allowances and, previously, mobile phone subscriptions, which can be bought and very quickly resold in two different EU countries.

---

[5] This involved either changing the VAT billing rules or simply eliminating VAT on cash transactions on CO_2 allowances.

Another fraudulent scheme needing to be stopped involved hackers penetrating accounts opened by companies so they could obtain their CO_2 allowances and immediately sell them, then vanish. In fact this is not unlike bank card fraud, whereby the intruder gets past the security software that normally allows only the cardholder to use the money in the account.

The first conclusion to be drawn is that in the launch stages of new environmental markets, regulating spot operations is as important as doing so for derivatives. From this standpoint, the weaknesses of the European system stemmed from a certain initial ingenuousness. In giving European citizens access to the carbon market, opening an account on the national registries was made too easy at the outset, inevitably attracting those who saw it as a way of making easy money. Remember that the cumulative value of CO_2 allowances amounts to several tens of billions of euros. Faced by the risk of misappropriation, the rules of access to the spot market had subsequently been greatly tightened up. Opening an account on a national register is now subject to restrictive conditions that limit access to the spot market.

But doing so does not settle the question of the overall monitoring of the market and of the specific role to be played by the public authority. The carbon market has its own specific feature that gives the regulator considerable scope for action: the supply is controlled by the public authority, who sets the emissions caps and the initial distribution of the supply through the allocation of allowances. Moving from allocating to auctioning allowances in the third stage ought to be the occasion to reinforce the regulation of this market.

Achieving this requires first setting up a team of high-level experts to gather information and analyze the fundamentals of the market. The European Commission does not have this type of expertise, which is more the domain of a kind of central bank. Let us call the derivatives market regulation body the carbon central bank. In supervising auctions, the carbon central bank would firstly control the quantity of CO_2 allowances placed on the market. This is the first prerogative of a central bank: creating money. In its monitoring role, it would then have to anticipate the risk of a dominant player or manipulation appearing in the market. It would also need to be able to intervene with supply and demand in order to smooth out erratic fluctuations of prices in the event of a crisis, averting the risk both of skyrocketing prices and of a price collapse. A final task would be to oversee the links between the European market and carbon markets operating elsewhere in the world. This is another prerogative of a central bank: watching over exchange rates. In the future, the connection between the EU ETS and other markets would allow trading to be extended and the influence of the carbon price to be strengthened.

Price formation

As in any market, the price of carbon at a given moment reflects the equilibrium between supply and demand. Supply is set by the public authority according to the environmental objectives aimed at. It is constituted by the emissions cap adopted, to which credits may possibly be added for emissions reductions obtained outside the system. It reflects the scarcity of the right to emit CO_2 into the atmosphere and is the market's most important compass. If the public authorities set emission caps sufficiently in advance, market players have long-term information and the market gains depth. In the American SO_2 market, for example, emissions caps are known thirty years ahead.

Demand depends on the amount of emissions coming from installations, which have to surrender the number of allowances corresponding to the CO_2 they have emitted. Part of this demand depends on parameters that are independent of companies' choices. Emissions are strongly influenced in the short term by weather conditions, which in turn affect the demand for energy. They are also dependent on the price of energy and the level of economic activity.

But the demand for allowances also depends on the possibility of reducing emissions on the part of the companies concerned. When the emissions reduction cost is lower than the price of carbon, companies have an interest in lowering their emissions in order to sell allowances (or to buy fewer of them if they are in a deficit situation). Overall demand on the market thus falls. In the reverse situation, companies gain financially by buying allowances rather than reducing their emissions. They consequently contribute to increasing the demand. If the market functions properly, companies therefore tend to equalize their marginal cost of emissions reductions with the price of carbon given by the market.

During the first period, the market equilibrium price showed considerable instability. Under the impact of demand from electricity companies, it rose sharply in 2005. In 2006, information made public by the European Commission revealed an overall excess in the supply of allowances over the demand from producers. As a result, the price fell, and then, logically enough, continued downward towards zero when it became apparent that this surplus supply would persist until the end of the period. The price instability of the first period was amplified by the 'no banking' rule forbidding participants to retain unused allowances from the first period for subsequent periods. Fortunately this rule, exclusive to the European market, has since been revoked and the functioning of the market has improved as a result.

During the second period, the supply of allowances was tightened, but this constraint was relaxed by the option given to producers of using up to 13 per cent Kyoto credits to meet compliance objectives. When the second period began, most analysts were expecting that, with economic growth, there would be greater upward pressure on prices. But the onset of the economic recession upset these forecasts and led to a fall in the carbon price during the winter of 2008–9. Compared to the first period, though, an upside factor was present, namely the possibility for producers to save unused allowances for subsequent periods. Because they expected a tightening on constraints for the third period, producers were motivated to keep their allowances for later use. As a result, the European allowance price was sustained and now reflects the expectation of a scarcity in carbon emission rights beyond 2012.

Two criticisms are often made of the European cap-and-trade system: too much price instability and too low a price level. Let us look at each of these in turn.

Producers need the market to send clear carbon price signals so that they can make their decisions. They cannot get this information if the price of carbon oscillates wildly, like foreign exchange rates or stock market prices during a financial crisis. During the first period the price fluctuated for institutional reasons, as we have seen. On the other hand, the CO_2 price in the second and third periods has not shown this volatility. It is, for example, considerably less volatile than gas and electricity traded on European markets, and somewhat less than oil products. It does, however, fluctuate more than the price of coal, which has been stable up until 2010 because of the stabilizing effect of long-term contracts. Contrary to what is often thought, carbon has thus become a commodity with low volatility in the markets.

Even if it fluctuates less, is the price of CO_2 nevertheless at 'the right level'? A recurrent criticism of this market is that the price is too low to trigger the investments needed for the transformation of the energy landscape. But is the market to blame? If it works properly, the carbon market should on the contrary lead to the lowest possible price given the cap that has been set. Its role is to minimize the economic cost of emissions reductions, not to increase them. If the price is considered to be too low, then it is the level of the cap that should be called into question, and the lack of resolve on the part of public authority to curb the right to emit.

Emissions reductions

The aim of the European carbon market is to reduce CO_2 emissions from European industry in 2020 to 21 per cent below their 2005 levels, or a 31 per cent reduction compared to 1990 levels. This objective can be partly attained through 'offsetting', i.e., by using carbon credits from projects that have reduced greenhouse gas emissions outside the system. The fact remains that this objective is ambitious and no-one should imagine that it can be attained by minor modifications in relation to past tendencies. The role of the carbon price is to encourage producers to make choices that diverge sharply from previous trends.

It is difficult to predict what type of emissions reductions companies will decide to make. Indeed, experience shows that the introduction of a carbon price generally results in emissions reductions that are quite different from those expected by experts or predicted by models. On the other hand it is instructive to analyze what happened in the trial period. Some of the earliest studies on the subject were written with great insight by Denny Ellerman, one of the first people to have examined emissions allowances systems, and the Austrian economist Barbara Büchner.[6]

Broadly speaking, we can say that the introduction of the EU ETS has led to a reduction in emissions of around 120 to 130 million tonnes of CO_2. Although this reduction is modest – between 2 per cent and 5 per cent of initial emissions in three years, not enough to put the energy system on a satisfactory long-term trajectory – it is a lot more than anything else that has been done. No other European measure in relation to the climate has achieved even a tenth of these figures.

For the most part, these reductions have been obtained in the fifteen countries of western Europe, whereas the constraint has been much less in the twelve accession countries of eastern Europe, which have benefited from more advantageous allocations and have been net sellers of allowances. In the fifteen-member EU, reductions have been particularly sizeable in Germany and the UK.

---

[6] Denny Ellerman has, in particular, produced the first comprehensive economic assessment of the SO_2 market in the US.

The electricity-generating sector has accounted for the largest part of emissions reductions. Two main levers have been used: fuel switching and gains in efficiency. The carbon price has first of all changed the way in which power stations are called on in the short term in response to demand from the grid. It has favoured low-emission and flexible producers (hydropower and gas) over coal-fired installations. This type of fuel switching has been obtained without any additional investment, simply by optimizing the management of existing installations. Furthermore, the carbon price has stimulated investment for raising the productivity of the least efficient coal-fired power stations.

A closer look reveals that industry outside the power sector has also changed some of its practices in response to the introduction of a carbon price, including acceleration of efficiency gains with modest investments – control and regulation systems, cogeneration – and sometimes fuel switching, for example, through the combustion of waste and the use of blast-furnace slag in cement production.

These reductions have been obtained from existing installations without major investment. Yet the construction of a low-carbon energy system requires that investment be drastically changed. Will the second and third stages of the European carbon market enable this new round of investment to be implemented? The economic and financial crisis had a much more determining effect than the price of carbon at the beginning of the second period. New energy investments were frozen and decisions were largely postponed. But a great number of options remain open, and the example of the Drax and Belchatow power stations suggests that investment decisions will increasingly take into account the price of carbon.

In 2008, Drax decided to invest £50 million so as to be able to use wood in addition to coal. Doing so will give it greater flexibility as well as having a significant effect on emissions. If it manages to substitute 10 per cent of its fuel, it will save something like 2 million tonnes of

CO_2 a year. With a CO_2 price at €20, the investment pays for itself in less than five years.

Belchatow has chosen to position itself in relation to carbon capture and storage. By 2014 a large-scale pilot scheme should allow more than 80 per cent of the emissions from the new 858 MW unit currently under construction on the site to be captured and stored. This way 1.8 million tonnes of CO_2 a year are expected to be returned beneath the ground starting in 2014. Depending on the results obtained, it will then be decided at what rate this technique will be put into general use. Financially, the current price of CO_2 is not high enough to make the investment cost-effective, but it is substantially contributing to this and has served as a catalyst for the operation.

Diffuse emissions: the market plus a carbon tax?

Even after the extending of the carbon market after 2013, various energy emissions, coming mainly from land transport and the use of buildings, will remain outside the emissions trading system. Yet these emissions are increasing more rapidly than those from industry. Why not incorporate them into the carbon market?

One could imagine all of us having our own CO_2 allowance capping the emissions linked to our day-to-day life requirements, such as transport, accommodation and food. Thrifty behaviour would enable allowances to be freed up and sold, while excessive use of air-conditioning or a liking for sports cars would oblige them to be bought. Some economists recommend such a system, with the issuing of carbon cards showing one's ongoing debit or credit situation as regards the right to emit CO_2.

There is a much simpler and less expensive way of extending the carbon market to diffuse sectors. One simply has to cap the carbon content linked to the use of fuels at the level of refineries or of points of entry into the network for imported gas and oil products. At the moment European refineries have to pay for the carbon given off by the processes of converting crude oil into refined products. But these emissions amount to less than 10 per cent of the emissions that will be given off when these products are used in cars or boilers. All that would be required to have complete coverage of diffuse uses would be to place the remaining 90 per cent within a cap-and-trade system. This upstream approach has not been adopted by Europe. To ensure that diffuse emissions carry a price signal, the market can be complemented by a carbon tax system – which is what Ireland and France tried to do in 2009.

Ireland, the 'Celtic tiger', which in a matter of a decade or two became one of the wealthiest economies in Europe, was hard hit by the economic and financial crisis. Like Sweden in the early 1990s, it has engaged in shock therapy, with a massive bail-out of the battered banking system, which has raised the public deficit to more than 10 per cent of gross domestic product (GDP), followed by a painful austerity programme. The carbon tax project introduced in January 2010 is one of the components of this therapy. It applies to all users of fossil fuels which are not in the EU ETS. Set at €15 a tonne, the rate is intended to reflect the price of CO_2 allowances on the market. It is not refundable, except for the poorest households, which have benefited from an increase in the allocation for combating energy insecurity. The tax does not seem to have had negative economic effects and is providing the government with the tax revenues it needs without adding to the corporate tax burden on companies that have invested in Ireland.

The French carbon tax was less successful. Launched in a blaze of publicity by the French president in September 2009, the project came up against a censure vote by the Constitutional Council in December and was quietly abandoned in March 2010. Apart from its legal weaknesses, the implementation of the project came to grief over the vagueness and ambiguities surrounding the corporate sector, the government having never clearly specified who should pay and who should be compensated in this area. Lacking any clear vision and authority, it was then a matter of negotiating, category by category, rates à la carte and exemptions on a production line basis. The project has since joined the ranks of those many reforms that are often announced but rarely implemented.

Despite French setbacks, the linking of national domestic taxes on energy emissions with a cap-and-trade system is a promising way forward. Its general adoption would enable the whole of the European energy landscape to be covered. There are a number of basic rules to be followed for it to succeed: exempt installations that have been allocated allowances; set the tax rate at a level that is not too far from the CO_2 allowances price so as to tend towards a single price throughout Europe; and have simple, transparent rules in regard to refunds that are stable over time. Setting up of such a decentralized system would, moreover, have a great institutional advantage in Europe: it would allow the sacrosanct unanimity rule in relation to tax decisions to be by-passed.

Carbon leakage and border adjustment

Doesn't part of the emissions reduction resulting from the introduction of a carbon price in Europe risk being illusory? The EU is an area open to the world. Yet the rest of the world has not set up an equivalent system of pricing carbon. There is therefore a risk that paying for carbon in Europe undermines the competitiveness of European firms which emit large amounts of CO_2 to the advantage of other businesses which emit just as much but do not pay for their emissions because they are not located in Europe. This is what is known as the risk of 'carbon leakage'.

As regards existing installations, Richard Baron, an economist at the IEA, has coordinated empirical studies on most of the sectors at risk. His conclusions are incontrovertible: there is not the slightest evidence of any carbon leakage during the launch period of the European carbon trading scheme, including such sectors as cement, steel and aluminium, where CO_2 accounts for a significant proportion of the production cost.[7]

But the constraint was reduced during the market's launch period. Is there not a risk of carbon leakage appearing as the constraint is gradually tightened? Once a cement plant or an aluminium foundry comes on stream, industrial logic leads it to maximize its use because of the relation between fixed costs and variable costs. The price of carbon is only one of the many costs that make up these variable costs. In addition, transport costs must be taken into account. In the case of cement, these costs break up the market into regional segments, and only port-based installations are directly subject to international competition. Except in exceptional and geographical limited cases, the price of carbon will not by itself redefine the geographical distribution of production from existing industrial installations.

The real carbon leakage question concerns investment decisions for new factories. In the context of the European emissions trading system, this question has up until now been pushed to one side because of the existence of national reserves enabling CO_2 allowances to be allocated free of charge to new installations. This type of incentive consists of subsidizing investment in greenhouse gas emitting factories with free allocated allowances. This practice is to be phased out as of 2013. Hence the risk of carbon leakage by the setting up of new factories outside Europe, where emitting carbon still costs nothing.

Some economists, such as Olivier Godard in France and Michael Grubb in the UK, who specialize in emissions trading schemes, recommend introducing mechanisms to neutralize these carbon leakage

[7] Aluminium production is carried out by electrolysis and does not directly emit CO_2. The process generates other greenhouse gases, which will be incorporated into the European emissions trading scheme as from 2013. On the other hand, it requires a large amount of electricity. The aluminium industry therefore bears the cost of carbon included in its electricity purchases.

risks at the borders (Godard 2008). The idea is to set up a kind of airlock on the EU's external borders with a dual corrective mechanism: imposing an import levy at the moment of entry into the EU proportional to the carbon content of the products imported, and refunding this same amount for products exported.[8] The carbon content of products should be calculated on the basis of the best technology used from an environmental standpoint. To produce a tonne of cement, between 600 and 1200 kilograms of CO_2 may be emitted. In this instance, the levy at the border should be calculated on the basis of 600 kilograms of CO_2 per tonne of cement, so as to protect only those European cement plants that are most efficient in terms of their emissions.

The great advantage of border adjustment measures is that the public authority recovers the extra revenue obtained by non-resident producers arising from not having paid for their emissions. Symmetrically, part of the revenue from levies received from imports should be used to finance the export of carbon-intensive products. This second element, which seems somewhat unnatural from the environmental angle, is essential if one wants to neutralize the effects of the carbon price on the conditions of competition. Depending on whether the sector subject to border adjustment is a net importer or exporter, it will therefore be a contributor or beneficiary from a financial standpoint.

The carbon levy or subsidy at the border is a powerful incentive to pass on the CO_2 price in industrial selling prices. No doubt this incentive will be used by producers subject to the cap, which are often in an oligopolistic position. The goods concerned are largely intermediate products consumed by other industries. The cost of border adjustment measures will be borne by the industries downstream

---

[8] This system is very similar to the one set up in the EU in 1962 in the context of the Common Agricultural Policy. The guaranteeing of internal agricultural prices is combined with a system of levies and refunds at EU borders.

which buy these intermediate goods. Border adjustment measures may encourage Arcelor-Mittal not to relocate its steel sheet operations. But it will encourage Renault and Volkswagen to manufacture their future models in Turkey or Ukraine, even if it means using sheet steel of European origin, the exports of which are subsidized through the carbon price.

The price signal sent by border adjustment measures thus risks being rather different from what its promoters are seeking: rather than protecting the whole European industrial fabric, it risks weakening industries downstream by protecting those upstream. What sort of political signal does that give?

Geographical widening

Despite all claims to the contrary, border adjustment measures risk being seen from outside Europe as sending a message of withdrawal behind its borders, or even of protectionism. Using it as a negotiation argument to persuade other countries to adopt carbon regulation and thereby escape the levy is not credible, for it overestimates Europe's weight in international climate negotiations.

An alternative measure would on the other hand be desirable within the perspective of geographical widening: incorporating maritime transport into the emissions trading system, similar to what is going to be implemented in relation to air transport. Every year the transport of merchandise by ship emits about 20 per cent more CO_2 than civil aviation. It is increasing these emissions twice as fast as total energy emissions due to the expansion of international trade and industrial relocation. Like air carriers, shipping companies benefit from advantageous tax regulations for their fuel purchases and they have no economic incentive to reduce their emissions. Technically, connecting these emissions to the European emissions trading system for traffic entering and leaving European ports would be no more complicated than for air transport. It would directly affect a quarter of the world's merchandise trade. As such it would be a first step towards applying carbon prices to a sector whose rates send an important signal as regards decisions about the location of industry and the organization of logistical chains.

But there is a more fundamental reason for enlarging the system: the price of carbon is based on the credibility of the public authority's commitments. For in an open global economy in which Europe emits less than 10 per cent of total energy emissions, its credibility will be eroded if its market does not quickly become part of a larger system for which it will have been the prototype. From this angle, the stakes are highest in regard to the US and the large emerging countries.

Two carbon markets are already in operation in the US: the Chicago Climate Exchange has since 2001 brought together a coalition of actors on a voluntary basis; and the Regional Greenhouse Gas Initiative (RGGI) is made up of ten states from the north-east of the US which have capped greenhouse gas emissions from their electricity generating plants until 2020. A third market is expected to open in California, the first US state to have capped its greenhouse gas emissions until 2050.

Innovative and interesting though they are, these initiatives do not measure up to what is at stake. For this reason the US Supreme Court has entered the fray. This venerable institution can hardly be accused of indulgence towards militant ecologists. Nevertheless, in April 2007 it set itself at odds with President Bush, announcing that it was the responsibility of the federal government to regulate greenhouse gas emissions in accordance with the Clean Air Act, in other words with the law as it stood.

This Supreme Court decree gives legal ammunition to President Obama to get a federal law regulating greenhouse gas emissions passed by Congress. The proposals, which have been under discussion since 2003, comprise a comprehensive federal cap-and-trade scheme that covers emissions from the energy sector, heavy industry and, in certain instances, transport and buildings.[9] They fall within a long-term perspective, with emission trajectories targeting 2050. Agriculture and forestry have not been capped, but could receive tradeable carbon offset credits on the market in the context of projects reducing their emissions.

The linkage between the future US federal market and the European system is likely to develop gradually. It is probable that the American market will operate initially with a weaker constraint, and therefore with a lower price per tonne of CO_2, until such time as the two markets are entirely fungible. But harmonizing the measuring and verification rules from the outset would enable indirect links to be quickly introduced through the project mechanisms. Such harmonization would strengthen the mutual credibility of the two systems and would be a large step forward in terms of setting up a global carbon price system for emissions arising from energy production.

The European cap-and-trade scheme: a prototype for a global system

With the CO_2 cap-and-trade system, the EU was able to set up the first multinational carbon pricing mechanism. The system brings together some thirty nationalities, each with different cultures and languages. It covers countries at different economic levels and with sometimes opposed energy orientations. Thanks to its flexibility, it has allowed the system to overcome short-term conflicts of interest. Its credibility is based on the capping of industrial CO_2 emissions by the European public authorities. It is robust: no European industrial energy actor

---

[9] In 2003 Senator McCain, the Republican presidential candidate defeated by Obama, sponsored the Climate Stewardship Act, the first legislative proposal for federal regulation of greenhouse gases. Since then, several dozen draft bills have been discussed in various committees in both the House of Representatives and the Senate.

imagines returning at some future point to a system where carbon is free.

This system is a prototype that Europeans constructed in just a few years to achieve common objectives in reducing greenhouse gas emissions. Its complete success now depends at an internal level on two key variables: the capacity of the carbon market to trigger the investments needed to build a low-carbon future; and the capacity to extend the price signal to diffuse energy emissions not directly covered by the scheme. These outcomes are still uncertain, and Europe has less chance of attaining them if it remains isolated.

On a global scale, the capping of European industrial emissions is small beer: a mere 4 per cent of world emissions are directly capped. But the real issue is now for Europe to put its prototype at the service of a wider system in which a growing proportion of industrial and energy emissions will be capped. The first stage will be an American federal system, of which the many proposals discussed in Congress show that it is likely to provide wider market coverage and greater temporal depth through objectives targeted in the long term. Subsequent stages will involve persuading the large emerging countries to join a common system, whose flexibility allows many compromises in terms of both the scope of the carbon constraint and the distribution of burden sharing.

At a technical level, the credibility of the European system is based on homogenous, reliable methods of calculation that allow one tonne of CO_2 from energy production to be converted into hard currency. Can such carbon accounting apply to emissions arising from agricultural and forestry that are not covered by the existing shared cap-and-trade system?

FIVE

Intensifying agriculture to safeguard forests

Marjohan proudly displays a wad of cash, more than 800,000 Indonesian rupees in freshly printed notes. It's his first pay packet, which he will hand over to his mother. He has been taken on to work for a palm oil operation, 35 kilometres from his village, set up in the middle of the jungle. His job involves cutting down and then burning the trees of the tropical forest. Once the forest has been cleared, he hopes to find a permanent job on the new plantation, which would enable him to free himself from the family rice-growing business.

With Marjohan's wages, the family will be able to buy a motor-bike. As well as helping them go back and forth to the nearby town Samarinda, he is also intending to use it to travel to work. At present he is dependent on the bus chartered by the Australian plantation manager, which does the return journey once a week. He hopes to have a little more freedom of movement once the family motorcycle has been acquired.

If we include the emissions produced by Marjohan's new job in the national inventory of Indonesian emissions, the picture is completely different. For each hectare of forest cleared, the combustion of the wood burned on the spot emits around 600 tonnes of carbon dioxide (CO_2) into the atmosphere. When the forest has grown in a damp environment such as a peat bog, the soil has to be drained before planting palm trees for the plantation. This operation gives rise to a second release of carbon that lasts several decades. Deforestation is by far and away the largest source of greenhouse gas emissions in a country such as Indonesia. If we include the effects of deforestation, Indonesian emissions leap from 2.5 tonnes to nearly 10 tonnes CO_2 equivalent per capita.

Emissions linked to agriculture and forestry are often presented separately from each other, as if forestry could be dealt with independently of agriculture. Marjohan's story reminds us that this makes hardly any sense. The reason why Marjohan is required to cut down the forest is to increase agricultural production (of palm oil). The area allotted to forests has always been dependent on how mankind manages its agricultural resources. In order to halt tropical deforestation on a long-term basis, it is therefore a matter of acting on its causes, which are largely agricultural.

Emissions from agriculture and forestry

Amounting to around 2 to 2.5 tonnes CO_2 equivalent per capita, a third of the world total, greenhouse gas emissions arising from agriculture and forestry fall into two main categories: those resulting from farming practices, which primarily concern methane and nitrous oxide; and CO_2 emissions caused by deforestation and soil drainage. These emissions cannot be calculated with the same degree of accuracy as emissions from energy production.

The three main types of emission associated with farming are nitrous oxide from the use of nitrogen fertilizer and methane given off by ruminants and from ricefields. In national inventories, these emissions are calculated by multiplying the area (or the number of animals) by standard emission factors. The distribution of agricultural areas (and of livestock) is relatively well known in developed countries where agricultural policies frequently grant subsidies on the basis of the number of hectares or heads of livestock. The use of satellite images allows the figures declared by farmers to be checked with great accuracy. On the other hand, very little is known about the distribution of agricultural land and livestock in developing countries. This is one problem. Another major problem is how to determine emission factors. When a farmer spreads a certain amount of fertilizer on a plot of land, the nitrous oxide emissions resulting from this may vary by a factor of ten depending on whether it is raining or sunny during the weeks that follow. A cow emits very different amounts of methane depending on what it eats. A ricefield's nitrous oxide emissions vary greatly depending on the method and duration of irrigation.

Knowledge of emissions linked to tropical deforestation comes up against the same kind of problem. The statistics on forest areas supplied by the Food and Agriculture Organization (FAO) are compilations of national data of greatly varying quality, since many countries do not carry out forest inventories from field data. Thus emission factors are rough approximations. The use of satellite imaging can extend our knowledge of carbon exchanges between forests and the atmosphere, and is currently being used in real time in Brazil, the only large forested country to possess an instrument for tracking deforestation. But satellites are not able to measure carbon emissions resulting from forest degradation or emissions arising from drainage or from ploughing up the soil after the trees have been cleared. Yet when forest soil dries out, the release of carbon can be ten times greater than the emissions previously given off by burning the trees.

The first consequence of these uncertainties concerns the overall statistics, which are highly unreliable and difficult to use.[1] As long as

---

[1] The fourth Intergovernmental Panel on Climate Change (IPCC) assessment report estimates, for example, that tropical deforestation releases 5.8 billion tonnes of CO_2 into the atmosphere plus another 1.9 billion from the draining of wetlands. The estimate of emissions from deforestation seems high, for it is based on FAO forest statistics that have since been revised. Analysis of satellite photos suggests that emissions linked to deforestation are around 3.8 million tonnes (excluding draining and soils). These estimates do not take account of emissions generated

it is a matter simply of producing statistical yearbooks, the implications are not too serious. But once this data is deployed for international negotiations, its unreliability becomes a major issue. The implications also matter at a microeconomic level: in the absence of reliable carbon accounting, it is impossible to put agricultural and forestry operations into a carbon trading system, where each business calculates and reports its emissions. If reductions in emissions from farming and forestry are to be efficiently and reliably implemented, it is therefore necessary to create instruments suited to the measurement and verification systems that can be set up.

Another important difference compared to emissions from energy production relates to geographical distribution. The proportion of emissions from agriculture varies a lot from country to country. In New Zealand for example, where there are ten times as many sheep as people, this represents nearly half the total, as against less than 10 per cent for developed countries as a whole. In developing countries, emissions arising from agriculture amount on average to 40 per cent of the emissions from energy production. This proportion rises to over 200 per cent in the least developed countries, whose economies are extremely undiversified. Since 1990, emissions from agriculture have slightly declined in industrialized countries, mainly because of fewer cattle and reduced inputs of nitrogen fertilizer. They have increased by a third in developing countries, which now account for more than two-thirds of emissions arising from farming. Emissions due to tropical deforestation are concentrated in some ten developing countries, including Brazil, Indonesia and the Democratic Republic of Congo.[2] The geography of farming and forestry emissions

by the degradation of tropical rainforest, which are hard to measure and vary considerably according to the area concerned. In the Congo basin, the largest area of rainforest after Amazonia, emissions linked to forest degradation are high, even though clearing remains limited due to the lack of security.

[2] Of the developed countries, only Australia and France (through its overseas territory Guyana) have a significant amount of tropical forest.

is therefore radically different from that of energy-related emissions, with developed countries accounting for a small and decreasing proportion of the former.

Agricultural and forestry carbon sinks

Another specific characteristic of the agroforestry system is that it can store carbon in plants and in the soil. We then speak of 'carbon sinks'. The agroforestry system as a whole is a vast carbon sink, with more carbon tied up in the soil and forests than in the planet's total oil deposits. The forests alone contain almost as much carbon as the Earth's atmosphere. These sinks can empty or fill up in response to human activity.

The storage of carbon by plants is not permanent. Carbon is imported as a result of photosynthesis, which enables plants to withdraw CO_2 from the atmosphere as they grow. Carbon is then exported back into the atmosphere during the decomposition or combustion of dead plants. Carbon sinks are formed when imports are larger than exports over a number of years. Trees are the main agent, though some farming practices can also contribute to the process.

A young forest stores carbon as it grows. Every year growing trees trap atmospheric carbon, acting as a carbon sink that at the end of a hundred years will contain around 300 tonnes of CO_2 per hectare if the forest is situated in the temperate zone. Once mature, a forest can live a very long time. The storing of carbon by living trees is offset by the emissions from dead trees. Carbon exchanges also depend on how the forest is managed: a forest that is becoming degraded is a net emitter of carbon. If a forest is cleared after a hundred years by burning the wood on the spot, the 300 tonnes of CO_2 per hectare returns immediately to the atmosphere. If felled trees are left to decompose, the process will take longer. And if the wood is used for building or furniture, the lifetime of carbon sinks is further prolonged. This is another way of retaining trapped atmospheric carbon. Thus you and

I contribute to the process, even if we are not aware of doing so. That antique dresser in your living room is a carbon sink, and should be taken good care of!

Primary tropical forests, which were not planted by man, have carbon exchanges close to equilibrium. When a hectare of one of these forests is suddenly cleared by burning, it immediately releases 500 tonnes or more of CO_2, that has been stored there for a considerable length of time.[3] Similarly, it takes around thirty years for a natural meadow or a crop on unploughed land to store carbon in the soil. By deep ploughing the land a few times after these thirty years, a farmer will have emptied most of this sink.

This reversibility of carbon storage by agricultural and forest sinks should be taken into consideration. A forest can indeed always burn, a hedge be cut down or a meadow ploughed up. Hence there is an obvious conclusion to be drawn: in the short term, the protection of existing carbon sinks benefits the climate far more than the creation of new sinks. Preventing the destruction of a hectare of tropical rainforest avoids emissions of 500 tonnes or more CO_2, whereas planting a hectare of new forest allows only a few tonnes a year to be laboriously gained. Planting trees is fine, but protecting those we have inherited is far better.

Historically, European and then North American expansion took place through clearing forests, which allowed agriculture and the raising of livestock to be extended. This process was reversed during the twentieth century, first in Europe and then in North America, with the expansion of boreal and temperate forests, thereby enabling carbon to be stored. On the other hand, Amazonia, the Congo basin and the tropical forests of Asia and Oceania are losing several million hectares every year, most of which go up in smoke so as to make way for agriculture and livestock.

---

[3] To simplify matters, we do not here take into account the soil of the forest floor which also stores and releases carbon.

Stopping this deforestation process would be highly beneficial for the climate. And doing so does not require the development of new technologies or investing in costly installations. It is for this reason that Nicholas Stern, and subsequently influential institutions such as the World Bank and McKinsey and Company, see it as a major emissions reduction opportunity at a low cost. The argument should be examined carefully. Defending the great forest basins against human predation involves finding satisfactory solutions to ensure food supplies without needing to eat into the forest. The operation is perhaps more complex than it may have appeared at first sight.

Economic crisis and food insecurity

The economic and financial crisis of 2008–9 revealed the fragility of the world food situation. Its initial victims were not the savers ruined by the stock market crash or the American families affected by the subprime crisis, but people suffering from hunger: some 100 to 150 million more between 2006 and 2009 (FAO, 2009; Shapouri *et al.*, 2009). The media have occasionally made reference to food riots in Indonesia, Egypt or Haiti. But they have remained silent in regard to the massive deterioration in the food situation since 2007.

The people affected are found to some extent in the outlying areas of cities. But mostly they live in the rural world, where poverty is worse than in cities and where families have often become net buyers (rather than sellers) of staple foods. Two factors have contributed to this large-scale return of food insecurity: the sudden rise in international agricultural prices from 2006 to 2008, which badly hit the purchasing power of families at the lower end of the income scale; and the economic recession of 2008–9, which directly reduced their resources. As a result, the number of people suffering from malnutrition has gone up from 850 million to a billion.

This crisis reveals two, more structural, phenomena. Between 1960 and 1990 the extent of hunger in the world was on the wane. From 1990, this progress ceased, despite the beginning of a demographic slowdown which in principle should have reduced pressure on resources. The number of people suffering from malnutrition stabilized at more than 800 million, mostly in Asia and sub-Saharan Africa. This halt is explained by factors both of supply and demand.

On the supply side, the first parameter to take into account is the slowing down in yield improvements per hectare, which has been observed in the agriculture of industrialized countries as well as that of developing countries. There are several reasons for this decline: greater competition in regard to water use, impoverishment of agricultural environments due to erosion, and insufficient investment and agronomic research. The declining rate of increase per hectare has not been fully offset by the extension of the area occupied by farmers, which amounts only to some 15 million hectares a year, very little in comparison to the 5 billion hectares in use worldwide. Relatively small though it is, this extension puts great pressure on fragile natural environments: savannah and semi-arid zones that are becoming desertified; and tropical forests that are being cleared. The extension of farming is the main cause of tropical deforestation.

On the demand side, the average quantity of food calories available per capita would largely cover the whole planet's food needs. In 2006, it was on average 2,600 calories per person in developing countries, while 2,000 calories is enough to cover an adult's needs. But the distribution of these calories among the Earth's inhabitants is distorted, with the spread of food models that are richer in animal products among the middle classes, while basic needs are not met at the lower end of the social scale. If greenhouse gas emissions from the agroforestry system are to be sustainably reduced, methods must be used that can at the same time improve the security of food supplies. Two kinds of levers can be operated: those relating to demand and those relating to supply.

The spread of new dietary models

Between now and 2050, changing demography will gradually ease the pressure on demand. Rather than the present 80 million more people a year to feed, it will fall to 55 million, and considerably less towards the end of the period, because the rate of increase of the world population is expected to fall from 1.5 per cent a year to less than 0.5 per cent by the middle of the century, with major geographical redistribution (in particular a declining Chinese population at the end of the period). But as we have just seen, as much as the number of mouths to feed, it is people's changing diet that weighs on food supplies.

The proportion of animal products is a defining characteristic of dietary models. From the standpoint of agricultural resources, the meat and milk we consume are vegetable products transformed by animals. This transformation has an energy cost: eating a kilogram of chicken is equivalent to eating 2.5 kilograms of grain. Indeed hens are excellent energy converters. But a kilogram of pork is equivalent to 4 or 5 kilograms of grain and a kilogram of beef to 7 kilograms of grain. The higher the proportion of animal products in people's food intake is, the greater the pressure on primary agricultural resources.

This proportion has stabilized in the food intake of the rich countries, at distinctly too high a level: food overconsumption in general, and of animal fats in particular, has become one of the main factors of morbidity and mortality.

Consumption of animal products is, on the other hand, growing very rapidly in emerging countries, as middle-class living standards rise. As a result, nutritional inequality in developing countries is becoming more marked. The additional calories used for meat production are no longer available for those suffering most from malnutrition, the number of which is again on the rise. At the same time, problems of overconsumption are beginning to appear. For example, the number of cases of diabetes in India is greater than in the US and one in every five Chinese adolescents is overweight.

These changes are far from trivial in terms of greenhouse gas emissions. Generally speaking, the ecological footprint of animal products is higher than that of plant products, since the former are obtained from the latter.[4] This is particularly the case for the carbon footprint of food products: eating a kilogram of chicken generates about as much CO_2 equivalent as eating 6 kilograms of bread. For steak, the figure is at least 20 kilograms.

Reorientating our dietary choices

In regard to changing the pattern of demand, one could imagine a generalized price on greenhouse gases that would influence household food choices by making the most carbon-emitting foodstuffs more expensive. To alter significantly our day-to-day expenditure on food, there would need to be a very high CO_2 price. Let us take two extreme cases: beef and oysters, for a carbon price set at, say, €20 per tonne of CO_2.

Because of the physiology of cattle, endowed by nature with four stomachs, beef is the foodstuff that contributes most to greenhouse gas emissions. It is estimated that on average the emissions produced by cattle, to which we add the emissions arising from transport and cold storage, are around 15 kilograms CO_2 equivalent per kilogram of beef consumed. If households are charged for this CO_2 on top of the price of the meat, it would add 30 centimes per kilo – an insignificant amount compared to the €15 a kilo one pays for a standard cut of beef in a supermarket. The price of carbon would barely appear on the label.

Although most of our foodstuffs contribute to greenhouse gas emissions, oysters on the other hand have the unusual property of

---

4 Ecological footprint is defined as the quantity of natural resources used to produce a given good. It is often expressed in global hectares. By extension we use the term 'carbon footprint' to refer to the greenhouse gas emissions content of products consumed.

storing CO_2. In an oyster weighing 100 grams, we eat around 20 grams of food and throw away the shell. This shell contains on average 35 grams of CO_2, which the oyster takes from the sea as it grows. Because oyster shells do not rot, the CO_2 stays there indefinitely. They are therefore a good carbon sink. Rather than buy a kilo of beef, we could shell out a bit more and buy 4 kilos of oysters, which would cost us €60 rather than €15. Would the carbon credits accruing to the storage of the CO_2 make up the difference? Our 4 kilos have trapped only 1.5 kilos of CO_2. So unfortunately our 3 centimes of carbon credits is not going to allow us to eat oysters more regularly.

The right way to change food choices is therefore through information and education. As a priority, we should target consumers in the rich countries, where diets have become too rich, especially in regard to meat products. Furthermore, this dietary model, extensively represented in the media and promoted by large advertising budgets, is tending to spread to the rest of the world. Formerly reserved for special occasions, beef is becoming an everyday foodstuff in dietary models of developed countries. Because of an overabundance of saturated fats, it leads to health problems such as high cholesterol, heart disease, obesity and so on. It also plays a large part in the highly undesirable development of mono-diets, exemplified by the hamburgers or kebabs served up by the million at local fast-food outlets, which have replaced the traditional greengrocer or fishmonger.

Must we then advocate dieting and the general adoption of vegetarian eating? Less restrictive menus have more chance of appealing to people. First of all, reducing beef consumption in favour of chicken or fish has a very beneficial effect on the carbon balance sheet. It should not be forgotten that a large proportion of the beef consumed is a by-product of milk, which has hard-to-replace nutritional qualities. Furthermore, leguminous plants such as lentils and beans are, like animal products, rich in protein. Including these regularly in one's diet provides the same nutritional value as meat. At the other end of the food chain, these plants have the property of storing nitrogen from the atmosphere. They can therefore be grown without fertilizer inputs and interspersed with other crops in rotation, thereby enabling carbon to be stored in the soil. This storage potential has nothing in common, in terms of size, with that obtained by increasing our consumption of shellfish.

World production of oysters is around 4.6 million tonnes a year. If we add in scallops, mussels and a number of other species, we arrive at a total world production of molluscs of 13.5 million tonnes. Using the same coefficients as for oysters, this tonnage sequesters almost 5 million tonnes of CO_2 a year, which will not return to the atmosphere for a long time. This amount is equivalent to about 10,000 hectares of tropical forest, less than 0.0005 per cent of the total area. Even if world consumption of molluscs were to increase tenfold, the effect would be tiny compared to the carbon storage potential of agriculture and forests.

New food-energy competition

Introducing the price of carbon into the household shopping basket is not the right way to re-orient people's dietary choices. So should we then abandon this kind of incentive for the agroforestry system? If climate policies generally adopt the pricing of carbon from energy production while leaving emissions from agriculture and forestry untouched, they introduce the risk of a new disequilibrium, caused by siphoning off agricultural products for energy use. This risk is also one of the lessons to be drawn from the imbalances that have appeared in agricultural markets over the last ten years.

During this period, biofuels were developed through generous subsidies by the US and Europe. If we take all these subsidies and estimate the number of tonnes of CO_2 saved by this new industry, we obtain an 'implicit' carbon price: one that would have induced investments in an economy operating without state aid, but with a carbon energy price in general use. The Organisation for Economic

Co-operation and Development (OECD) puts this price at several hundred euros for the biofuel industries developed in Europe and the US.

Between 2002 and 2007, the proportion of the US maize harvest removed from food use to produce biofuels rose from 10 per cent to 25 per cent. Overall, the US Department of Agriculture estimates that biofuel needs accounted for a quarter of the growth in world demand for grains and oilseeds during this period. The result was an unprecedented rise in the price of basic food products such as wheat, rice, maize and soya.

These price increases have resulted in a large number of people in poor countries experiencing food shortages, despite the safety nets that many governments have tried to set up in order to decouple internal prices from those of the world market. They have also led to an acceleration in tropical deforestation. Were it not for soaring world demand for oilseeds driven by biofuel needs, Marjohan would probably still be working in the family ricefields instead of being taken on by the palm-oil plantation set up with Australian capital.

If pricing carbon in the energy system becomes general practice with no safety net in the agroforestry system, pressure on energy demand for agricultural products is dangerously increased, with all the risks thereby entailed for food security on the one hand and protection of forests on the other. The carbon price acts as a subsidy favouring the competitiveness of biomass in the face of energy from fossil fuels and from which food uses do not benefit. A twofold lesson must be drawn from this. First, in the energy carbon markets, it is essential that accounting rules for carbon coming from biomass guard against the risk of new competition between energy demand and food demand. Second, ways must also be found of using the carbon price to stimulate the reorientation of agricultural supply towards methods that can both increase yields per hectare and reduce emissions.

The impact of climate change on agricultural yields

From the 1960s up until the mid-1980s, increasing agricultural yields allowed the food needs of a rapidly growing population to be met. In industrialized countries, these gains were achieved through a combination of genetic advances together with large-scale use of chemical products to protect potentially high yields but generally less robust crop varieties. In developing countries, the so-called 'green revolution' resulted in the selection of seeds suited to harsher conditions, combined with growing use of water and chemical products.

The slowdown in yield gains since the end of the 1980s partly reflects the reduced investment and research effort, but it also stems from the limits of an intensification model overly based on specialized and industrialized agricultural methods. It is for this reason that the medium-term projections by the FAO and the US Department of Agriculture assume that the slowdown in yield gains per hectare will continue, reflecting the exhaustion of the beneficial effects of these models. Neither of these organizations specifically incorporates the impact of climate change on future agricultural yields.

In view of the present dynamics of greenhouse gas emissions, it is reasonable to suppose that temperatures will rise by between 1°C and 2°C by 2060. If large changes in emission trajectories occur in the coming decades, the beneficial effects on the climate will appear only during the second half of the century, given the inertia of the climate system. According to the experts, warming to this extent would have an overall positive impact on the world supply of agricultural products, though with large-scale geographical redistribution effects. Moderate warming is beneficial for temperate zone farming in high latitudes where the range and yield of crops and livestock are limited by the cold. But at the same time it adversely affects the food supply potential of regions located at lower latitudes. The fall in potential supply will be particularly marked in dry regions of sub-Saharan Africa, Mediterranean countries and central and southern

Asia. The main cause of this will be the increasing scarcity of fresh water. Moreover, it is likely that the greater variability of the climate, particularly the alternation of droughts and floods, will accentuate the fragility of these countries' food supply.

Abundance in the global North, shortages in the South: the spontaneous adjustment to this shift will naturally involve greater recourse to international trade. Yet such an emollient supposition demands ill-considered confidence in the virtues of the market. The truth is that massive further investment in agricultural development in the countries of the global South will be required, giving overriding priority to the factor of production that is at risk of becoming the most scarce, namely water. And to avoid falling back into past errors, more diversified models must be tested, that conserve natural resources and emit less greenhouse gas.

Reducing agricultural emissions: an obstacle race

The reduction of agricultural emissions is by no means simple. One major difficulty is the fragmented and complex nature of farming. The development of emissions reduction projects entails bringing together producers and forming effective relay systems. In the industrialized countries, the cooperative movement has generally been the foundation on which a network of agricultural development institutions has been built. Such a network is particularly dense in a country such as France, with its cooperatives, chambers of agriculture, agricultural management centres and technical institutes. The majority of agricultural producers have access to these and regularly appoint their representatives during elections at chambers of agriculture. Institutions of this kind are much less developed in most countries of the global South, where a large number of small farmers are left to fend for themselves, if they are not simply repudiated by official organizations.

The cost of setting up such institutions is high and sometimes prohibitive. This cost is not generally taken into account in studies that compare the costs of reducing greenhouse gas emissions from agriculture and forestry to those of other sectors, and conclude that sources of high emissions reduction at low cost exist in the agroforestry sector. If this were the case, it is likely that many more emissions reduction initiatives would have already been launched. To exploit the potential for agricultural emissions reduction, one must first remove the barriers to entry by setting up suitable operational organizations.

In the short term, the emissions reduction potential in relation to livestock is a matter first of all of manure management from intensive livestock farming. Methane given off by livestock accounts for 7 per cent of the world's agricultural emissions. Manure management offers advantageous possibilities for methane recovery at a low cost. It also brings significant secondary benefits by reducing subsoil and olfactory pollution. There are currently nearly a hundred projects promoting manure management in developing countries through the Clean Development Mechanism introduced by the Kyoto Protocol. It is a shame that this is so little practised in the leading intensive rearing zone in Europe, namely Brittany.

The raising of cattle and sheep accounts for a third of agricultural emissions. In France, it is estimated that the cattle population emits twice as much greenhouse gas as the country's twelve oil refineries. Indeed the digestive systems of ruminants are veritable methane factories. These emissions are calculated using fixed emission factors for each category of animal. Yet the actual emissions depend on what the animals are given to eat: if one puts linseed, for example, in the manger, a cow's methane emissions can be reduced by a third. Here there is possibly room for considerable progress, but nevertheless we must be on the watch for carbon leakage. For instance, a cow fed on grass emits more greenhouse gas, but a permanent pasture stores more carbon than a cultivated field producing grain. The most reliable way of reducing emissions linked to the raising of ruminants is simply to have fewer of them, and this can be done by cutting down on their overall contribution to food supplies.

Rice growing requires a lot of water. The most generally practised method involves flooding the fields. As in wetlands, this gives rise to the anaerobic fermentation of organic matter, which in turn produces methane emissions, amounting to 11 per cent of global agricultural emissions. Advances in rice-growing techniques can help reduce these emissions, for example by changing the irrigation method so that water is conveyed to the plants without flooding the field or by introducing varieties adapted to growing in non-irrigated soil. Pluvial rice cultivation, for example, is being developed in tropical Africa, though with much lower yields than those obtained through flooding. We should be wary of the large-scale adoption of uncertain experimental techniques. Rice is grown by more than 200 million farmers throughout the world, who have been able to adapt to natural conditions. They produce and market the world's most abundant crop: rice is the staple food for half the planet's population. It is important not to take the slightest risk of yield loss on a product that is essential for food security.

Nearly 40 per cent of agricultural emissions take the form of nitrous oxide (N_2O) from the soil. By far the largest proportion of these N_2O emissions results from the use of nitrogen fertilizers, with most of the rest coming from manuring with organic matter. Nitrous oxide emissions are high in areas of intensive agriculture and are rapidly increasing in Asia.

To reduce these emissions, more than anything else we need to use less nitrogen fertilizer. The right approach here is to reduce inputs by using them more efficiently. But results are not guaranteed. When fertilizer is used cautiously, it is impossible exactly to match the quantity of input with the plants' growth needs. But carbon prices can be used to encourage farmers in intensive agriculture zones to use fertilizer more sparingly. To this end, offsets should be given to programmatic projects operating on a sufficiently large geographical scale to be able to calculate the emissions reductions obtained on the basis of statistical methods. Concretely, this type of project could be judiciously implemented with farm supply cooperatives which, in the absence of carbon offsets would be hardly likely to praise the merits of limited fertilizer use to their members: such practices reduce turnover!

Some people advocate the general adoption of less chemicals and more extensive agriculture, even if it means accepting lower yields. The right approach is rather to aim for agriculture that is less chemical but just as intensive, or even more intensive when agronomics permits. Having lower yields per hectare simply leads to more hectares being brought into use to make up the difference, which shifts emissions without reducing them. Organic agriculture is not an unequivocal answer either. The replacement of chemical fertilizers by organic fertilizers does not eliminate nitrous oxide emissions, and in some cases it can even increase them. On the other hand, if well managed, organic agriculture enriches the soil with organic matter, which enables carbon to be stored.

Replenishing agricultural carbon sinks

Agricultural carbon sinks consist of organic matter present in the soil (humus, micro-organisms, earthworms, etc.) and carbon stored in plants. These reservoirs can be filled or emptied, depending on farmers' practices. For example, deep ploughing or draining wetlands returns carbon to the atmosphere that was previously trapped in the ground. Conversely, building up a layer of humus in the soil by burying plant residues and carrying out suitable crop rotation, burning these residues in the soil to produce biochar, planting a hedge or an orchard, reconverting a field to permanent pastureland or cultivating without ploughing, are all practices that replenish carbon sinks.[5]

---

[5] Biochar is charcoal obtained by burning agricultural residues found in the soil. This age-old practice in Amazonia can under certain conditions store carbon in the soil by improving its texture and fertility.

There are permanently a vast number of carbon exchange flows between agricultural soils and plants on the one hand and the atmosphere on the other. The fourth IPCC report estimates that agricultural carbon sinks release around 150 million tonnes of CO_2 a year, a relatively modest figure. It also identifies a number of levers that would enable farmers to store more carbon. Very often, replenishing carbon sinks brings a great many secondary benefits, since it helps combat erosion and environmental degradation. One simply has to observe carefully various farming environments to be convinced that this is so.

Let us first consider the grain-growing plains to the south of Paris known as the Beauce. The landscape is flat and uniform and, on rainy autumn evenings with the wind blowing, rather dreary. Over the generations, farmers have removed all the walls, bushes, trees, hedges, etc. By doing so they have saved a little time manoeuvring agricultural machinery, and incidentally have slightly reduced their CO_2 emissions. But the ecosystem has been impoverished and the reduced plant cover means that more carbon is exported into the atmosphere.

Replanting hedges would bring numerous benefits. Ten kilometres of hedgerow enables 6 tonnes of CO_2 to be stored per year over thirty years. Some tens of thousands of kilometres of hedgerow could be replanted to good effect in the Beauce, bearing in mind the 1.5 million kilometres that have been destroyed throughout France since 1950. Hedges also add to the diversity of ecosystems that have become uniform and provide useful protection against soil erosion. The restoration of plant cover is a good way of replenishing carbon sinks.

Let us now leap over the Mediterranean, cross the Mitidja plain with its orange trees that surrounds Algiers and traverse the high plateaus that separate the Tell from the Algerian Sahara. We arrive at one of the world's largest arid zones. To the north of the Sahara are the high plateaus that extend from the Moroccan Atlas to the Nile plain. To the south lies the African Sahel. On the Algerian side, the farmers practise 'dry farming', inherited from the colonial period. Grain monoculture is standard. The soil is always deep ploughed to bury the chemical fertilizers. But the ground also has regularly to be left fallow to maintain yields. The landscape is bare and rock-strewn, with no remaining plant cover. Any hills are deeply furrowed by erosion. The soil has been depleted of its organic matter and increasingly resembles the stony ground of the desert.

If desertification is to be avoided, the plant cover must be restored and farming practices altered. But hedges in this environment would no longer suffice. What is required is investment in genuine forest plantations on all the slopes to halt erosion. And wherever possible, crop rotation should be introduced to break with monoculture, and especially to protect the soil against overgrazing by livestock.[6] The growing scarcity of water, which is indispensable for anti-erosion plantations, is a major obstacle. By stopping the advance of the desert, such investment will also reverse carbon leakage and will gradually restore the storage capacities of agricultural carbon sinks.

Let us continue southwards, flying over the desert, crossing the Sahel, until we come to Cameroon. As we gradually leave behind the Sahelian north and enter the savannah, the vegetation becomes rather more abundant, but is increasingly less resistant to the ravages of bush fires. Most of these fires are of human origin. Farmers and herdsmen here deliberately set fire to the natural vegetation, which enables them to plant crops for a few years or to obtain regrowth of grasses, thus avoiding having to move their herds thousands of kilometres. But this practice impoverishes the soil, which gradually loses its organic matter.

[6] In traditional forms of animal husbandry on the high plateaus, the livestock is adapted to the environment by two regulatory mechanisms: keeping the flocks on the move, a practice that is inherent in nomadism, and regulating their numbers by means of the high mortality rates in dry years. Since 1970, Algeria has imported fodder on a large scale, which has encouraged herdsmen to settle and has led to an increase in the number of animals, which to a large extent cannot be supported by the ecosystem.

To restore the storage capacity of soil and increase plant cover, agronomists recommend the same type of solution: intensify land development by substituting crop rotation for nomadic farming and by fertilizing the soil; and alter livestock raising practices so as to have sustainable forage resources. But the main problem lies elsewhere. It is finding ways of acting on the complex social structures that lead to these kinds of predatory developments.

Let us continue further south. We come to the tropical rainforest of eastern Yaoundé. The forest is a patchwork. Parts of the forest have been completely razed, cleared when the price of coffee skyrocketed ten years ago. Some of these are still being exploited, while others have been left fallow. Elsewhere the forest has been thinned by families who grow sweet potatoes and manioc. Let us go further in. The forest becomes denser and less light penetrates the canopy. Many of the trees have been cut down. Abandoned plastic bags reveal the presence of humans. Contraband timber sells well and a lot of it is removed by villagers for domestic use. As agriculture progressively establishes itself here in place of the forest, it releases CO_2, with each hectare cleared giving off several hundred tonnes of the gas. It is very hard to put a figure on the emissions resulting from the degradation of the natural forest.

The dangers of deforestation

Tropical forests take up around a third of the world's forested area, or 12 per cent of the total land area. They can be broken down into primary rainforests, with a high tree density, where the canopy forms a continuous layer largely impervious to light, and dry forests or tree-covered savannahs, where the tree density is lower and the cover discontinuous. Tropical rainforests are an incredible reservoir of living species and a huge carbon sink: each hectare of rainforest stores on average more than 500 tonnes of CO_2. Savannahs are a far less rich environment with very variable carbon storage per hectare, but generally no higher than 200 tonnes of CO_2. Tropical rainforests comprise three main areas, each with its own distinctive characteristics.

Because of the size of the Amazonian forest, Latin America accounts for more than half of the world's tropical forest. Its rainforest covers nearly 700 million hectares, or twelve times the area of France. One hundred million people live in this zone. More than 70 per cent of them live at least eight hours away from towns, in zones where there are conflicting possible uses for the forest. The population becomes sparser the further one goes into the forest: only four million people live in the 450 million hectares of deep forest. For the most part they belong to indigenous communities that have lived in the forest since time immemorial and have been able to adapt to it.

More than half of Latin America's rainforest is found in Brazil, which has set up a system for observing how the forest evolves and for protecting it. This system enabled the rate of deforestation to be kept down to around 1.7 million hectares a year over the period 1990–2005. The rate has increased since 2007 in response to the rise in international agricultural prices. The primary cause of deforestation is the growth of cattle ranches for beef production. Small-scale family farming, producing food crops such as manioc, also contributes to deforestation, as does soybean production, which eats into the dry forest, the *cerrados*, situated to the south of the Amazonian rainforest in the states of Mato Grosso and Para. The growth of sugar cane production does not directly encroach on the forest. But the agricultural structures are complex, and small farmers who are sometimes dislodged by new sugar cane plantations may migrate towards forested areas.

In Africa, three-quarters of the tropical forest is savannah and the remaining quarter is the second largest area of rainforest in the world, the massive Congo basin covering nearly 200 million hectares. This expanse of rainforest lies mostly in the Democratic Republic of Congo, a country of 60 million inhabitants confronted by armed conflict and growing poverty. The forest here is more densely populated than Amazonia: more than 100 million people live in the rainforest area and 250 million in the dry forest zones. This population is growing rapidly. Here poverty is the main source of predation suffered by the forest areas, with the encroachment of smallholdings cultivating manioc and sweet potato. In the savannah zone, the sedentarization of hitherto nomadic herdsmen is upsetting the traditional balance between itinerant herds and the natural environment. The main energy source for the local population is wood burned as fuel. Cash crops – cotton in the savannah, coffee in the rainforest zone – are a source of monetary income. The logging of high-value hardwoods such as mahogany is practised in West Africa, at a high ecological cost. Estimation of the area of forest destroyed or degraded is very uncertain in the absence of an observation system such as the one available in Amazonia. The loss of forest is probably in the order of 1.5 million hectares a year in the savannah zone and 0.8 million hectares for the Congo basin. Areas suffering from degradation are no doubt considerably larger.

In Asia, the forest situation is very mixed. China (which is not in the tropical zone) and India (which is partially) both engage in replanting, but China imports large quantities of illegally felled timber. Asian rainforest covers just over 210 million hectares, and extends over the south of the continent along an arc running from Burma, through Indochina and the Pacific archipelagos, to Papua New Guinea in Oceania. Half of the forest expanse is situated in the Indonesian archipelago. Part of this forest grows on peat bog, a second very important carbon sink beneath the trees themselves. This area is much more densely populated than Amazonian or the Congo basin rainforest: 450 million people live there. Tropical Asia has nearly 40 million hectares of plantations, with the main products being vegetable oil, rubber and tea. Every year, nearly two million hectares of rainforest disappear. The main causes of deforestation are the expansion of commercial crops such as palm oil, the extension of small-scale food production, and the logging of wood for export. It is in Asia that tropical rainforests are shrinking fastest. According to the most recent estimates, the CO_2 released from Asian forests and soils now equals that from the whole of Latin America.

Can putting a price on carbon save the tropical forests?

Halting deforestation will not be achieved by posting guards or militant ecologists at the forest edge. It is necessary to address the causes, which are largely related to agriculture and primarily concern the 900 million people who live in or in the immediate vicinity of forested areas. The means to be deployed vary considerably depending on the area concerned, and the role of carbon pricing would be very different from one to the next.

Consider the expansion of beef production on commercial cattle ranches in Brazil. Cleared forest rarely brings in more than €300 a hectare, since the grazing conditions are poor and the cattle density low.[7] Raising cattle on these new ranches first involves cutting down the forest, which contains around 600 tonnes of CO_2 per hectare. If the ranch owner were obliged to pay for carbon emissions at the price of €20 a tonne, it would cost him €12,000 per hectare. Investing €12,000 for an annual return of €300? Hardly! If it were applied in this type of situation, carbon pricing would be a sure-fire way of preventing tropical deforestation. Even at a quarter of the European price, the impact would still be very much a deterrent.

Marjohan's situation is very different. The tropical forest has been cleared and replaced by a plantation that will produce bioenergy: after the five years needed for them to become productive, the palm trees will provide oil that will then be transformed into biofuels. The 600 tonnes of CO_2 emitted into the atmosphere when

---

[7] Kenneth M. Chomitz (2006) shows that part of the deforestation occurs on land where the yields turn out to be low. For ranches, the figure he gives is $200 average income per hectare cleared.

the forest was cleared will eventually be recovered. Firstly, the trees in the new plantation will store up to 150 tonnes of CO_2 per hectare. The oil they produce will then help conserve CO_2 by substituting biofuel for conventional fossil fuels. Depending on the circumstances, it will require thirty to more than a hundred years for the new plantation's carbon balance sheet to become positive. Purely from the climate angle, should this reconversion be implemented or not? It all depends on the value placed on time: the destruction of tropical carbon sinks has an immediate cost, but generates long-term benefits from the standpoint of the overall CO_2 exchange with the atmosphere.

In many cases, particularly in Africa, deforestation results from survival behaviour – finding the wherewithal for heating or food – which has little to do with the price of carbon. It is therefore necessary to apply other levers that are concerned primarily with development. On the other hand, the carbon economy can free up financial resources allocated to such development projects.

These three scenarios show that the carbon price can play a very different role according to the situation. Sometimes it is useless. In other situations, as in the case of cattle ranches, or more generally extensive agriculture with high greenhouse gas emissions, it is a very strong incentive to protect forest sinks. But pricing carbon is not an unqualified panacea in cases where one wants to use natural tropical forest to produce energy. The non-energy value of such a forest, particularly its biodiversity, is not included in the price of CO_2. To assign a value to the other services provided by forests, such as biodiversity, different economic instruments must be used. Here lies one of the main limits of the economics of carbon sinks, which is based solely on the benefits accruing to the climate.

Carbon sinks can provide other services that are indirectly of value, such as maintaining the fertility of the soil for farmers, protecting landscapes that will benefit providers of tourist services or, in the case of mangrove forests, preventing the salination of fresh water and protecting against tidal surges. Finally, carbon sinks of the primary tropical forests provide services – such as the protection of biodiversity – that cannot be directly priced. This type of service is what economists call a 'public good': primary tropical forests no longer exist only in Brazil, the Democratic Republic of Congo, Indonesia and a few other countries, for it is all of humanity which benefits from the unmatched reservoir of living species comprising it. Using it for energy purposes by sending Marjohan to cut it down and convert it into a palm-oil plantation is a choice that either takes no account of the non-energy value of this public good or implicitly puts a low price on it.

Tropical forests in international negotiations

The United Nations (UN) discussions on incentives to halt deforestation were launched in December 2005 at the Montreal Climate Conference by a group of forest countries led by Papua New Guinea. The idea was to set up a system of economic incentives to halt the clearing of tropical forest on a sufficiently large geographical scale to avert carbon leakage. Since then it has come a long way and its principles were widely accepted at the 2009 Copenhagen Climate Conference.

The planned mechanism, commonly known as Reducing Emissions from Deforestation and Forest Degradation (REDD), involves giving credits at a national level for emissions avoided in relation to a benchmark scenario. Suppose, for example, that a country has 100 million hectares of rainforest that each year is losing 500,000 hectares and thereby emits 250 million tonnes of CO_2. Let us take these 500,000 hectares as the benchmark scenario. If the country manages to reduce forest loss to 300,000 hectares, it saves 100 million tonnes of CO_2 in relation to the benchmark scenario. These 100 million tonnes of CO_2 can then be credited. If the corresponding carbon credit sells at €20, the country receives a cheque for €2 billion. If it cuts down 500,000 hectares, the level of the benchmark scenario, it obviously has no right to any carbon credit. So what happens if it cuts down 700,000 hectares of forest?

This question is one of the most crucial aspects of the negotiations, and refers to the issue of the ever-possible reversibility of actions in relation to forest carbon storage. Until such time as this issue is clearly and transparently resolved, there is considerable risk that forest credits will be too generously distributed and therefore will not find buyers.

If the system is to work efficiently, it is not enough for the forest countries simply to acquire carbon offsets. It must also provide strong incentives at a local level, through projects that offer economically advantageous and sustainable alternatives to the farmers and stock-breeders who are currently induced to cut down the trees. It is at this microeconomic level, quite remote from the principles debated in international negotiations, that these alternative models for initiatives likely to halt future tropical deforestation will be tested. One of the conditions for their large-scale deployment is that ways must be found of transferring the value of carbon of the emissions avoided to the basic producers.

A long-term undertaking

The economic and financial crisis revealed the fragility of food supplies and the lack of concern of development policies that have disinvested in agriculture and support for natural resources. As a result, very heavy pressure is being exerted on the biosphere's carbon stocks, particularly in relation to tropical forests.

The reorientations to be undertaken by the agroforestry system to improve food security while reducing emissions will call for a major investment effort, in particular to enable farmers in subtropical zones to avert the risk of falling yields resulting from future global warming. To be effective, this effort should be accompanied by the spread of new, more diversified production models that use natural resources more efficiently. The spread of new techniques among the millions of basic economic units has a cost that is not easily comparable with the cost of developing capital-intensive technologies deployed in the energy sector. But it is evident that it will be a long-term undertaking, and that the cost will be much higher than is generally expected.

As has been shown by the example of first generation biofuels, pressure on resources will be increased by the general adoption of the carbon energy price, which acts as a subsidy benefiting bio-energy to the detriment of fossil fuels, and risks diverting a growing proportion of agricultural resources away from food production. In theory, the way to counter this situation would be to integrate agriculture and forestry into a shared carbon price system with the energy actors. But such integration is not possible in the absence of reliable carbon accounting for the agro-forestry system and because of the latter's highly fragmented nature.

Another route, that of the Kyoto project-based mechanisms, should therefore be explored. Through these, farmers are financially encouraged to reconcile higher yields per hectare with the protection and replenishing of existing carbon stocks. The proposal by the group of forest countries to finance non-deforestation with carbon offsets is one step in this direction. However, it cannot be deployed on a large scale as long as two key aspects of the system remain blurred: the financial management of the risk of the reversibility of carbon storage obtained by non-deforestation; and the practical use of the carbon price to change the behaviour of actors who are today motivated to clear the forests.

In parallel with the re-orientation of supply, acting on demand would help relieve the pressure exerted on natural resources by food production. The dietary model of Western countries, overly rich in meat and fats, is tending rapidly to spread to the middle classes of emerging societies. This model is far from optimal from a nutritional standpoint. No less than in the case of energy, this extravagant, high carbon footprint model cannot be generally adopted in a world whose finite resources must be better protected.

SIX

The price of carbon:
the economics of projects

The village of Bagepalli is located in Karnataka, one of the states of southern India. The area is semi-arid. The inhabitants collect wood and plant residues for cooking their food and boiling water to make it safe to drink. This requires increasingly long journeys. Indeed the resource is disappearing: 75 per cent of the forest in Karnataka is no longer being replanted. The method of cooking using traditional baked clay ovens is relatively inefficient, and the resulting household waste is toxic.

The local government is trying to help the villagers by remedying the situation as quickly as possible. It subsidizes the use of kerosene, which most families cannot otherwise afford to buy. But this type of measure makes the population dependent on public monies without providing a sustainable economic answer. In health and environmental terms, it is frankly harmful. Kerosene exposes the inhabitants to more toxic residues than wood and emits more carbon dioxide (CO_2) of fossil origin.

The local non-governmental organization (NGO), Women for Development, has worked for several years with the families from the village. In 2006, it was able to increase the scale of its work, thanks to the Clean Development Mechanism (CDM) introduced by the

Kyoto Protocol. The mechanism allows carbon credits to be given to projects that reduce greenhouse gas emissions. Collecting wood from a deteriorating environment and the use of subsidized kerosene results in each family emitting on average 3.5 tonnes CO_2 equivalent a year. In Bagepalli, the average family owns four cows. Processing the cattle dung and household residues in individual methanizers allows a substitute energy – biogas – to be produced. As its name suggests, this gas is of biological origin and comes from the fermentation of organic matter in the methanizer. Unlike natural gas of fossil origin, its combustion does not add any CO_2 to the atmosphere: if they were not processed in the methanizer, dung and other waste would emit the same amount of CO_2 during their natural decomposition.

The Bagepalli project, carried out in partnership with the Indian subsidiary of the French firm Velcan, should enable 5,500 families to be equipped with individual methanizers for producing enough biogas to meet their needs. These families will thus have access to an energy source that does not emit greenhouse gas and is much safer for their health. The implementation of the project is expected to reduce emissions by nearly 20,000 tonnes CO_2 equivalent a year. This reduction will be converted into that many carbon credits. Selling these carbon credits should cover the remaining part of the financing of the project, which would not be possible without this additional source of money, because the villagers, who have built their methanizers themselves, do not have the wherewithal to buy the materials and products required. Without carbon credits, the Bagepalli project would never have got off the ground.

Bagepalli belongs to the category of small projects developed within the framework of the CDM. It has a relatively small effect on emissions reduction but has a large impact on local development. Let us now move up to the industrial scale.

Pusan is the economic powerhouse of South Korea. A large proportion of the country's exports go through its port installations. A petro-chemical complex is located a few kilometres north-east of the city, in the port of Osan. One of the four adipic acid factories in Asia is found there, and every year it produces up to 150,000 tonnes of this acid, the main constituent of nylon. The standard manufacturing process emits substantial amounts of nitrous oxide: on average, more than 9 million tonnes CO_2 equivalent enters the atmosphere every year.

The Osan plant is owned by the French company Rhodia. The group is accustomed to greenhouse gas emissions and since 1993 has been keeping a record of its emissions. In 1998, it developed a technology using heat that destroys 99 per cent of the nitrous oxide emissions at its Chalampé site in France. This technology involves an initial investment on which a return can then be quickly secured if emissions reductions are priced. In 2006, the group decided to deploy the technology on the Osan site. In exchange, it should receive more than 9 million carbon credits a year until 2012. These credits cannot be used in Korea, where industrial companies are not subject to emissions regulation, but can be sold on the international market. It was this prospective return that induced the company to transfer the technology previously developed in France. Its social consequences are in no way comparable to those of the Bagepalli project. On the other hand its impact on global warming is substantially higher. In fact, if we calculate the figures, we can see that it would need 470 Bagepallis to equal the emissions reduction at Osan.

What is carbon offsetting?

The projects launched under the CDM result in what is known as 'carbon offsetting'. Neither the villagers of Bagepalli nor the Korean industrialists are obliged to reduce their emissions of greenhouse gases into the atmosphere. Other actors, such as European companies subject to emissions allowances, need to reduce their emissions to be in compliance with their cap. With carbon offsetting, these industrial actors are authorized to use carbon credits such as those accruing to the Bagepalli or Osan projects to ensure their compliance. The companies concerned have not reduced their own emissions. Instead they have offset them by buying the credits of other actors who are outside the cap-and-trade system. Each credit purchased is a certificate attesting to the reduction of a tonne of CO_2 emissions through a project that has been implemented.

Thus the two mechanisms complement each other. Cap-and-trade schemes create scarcity by capping emissions. This scarcity enables a carbon price to emerge, which is higher or lower depending on the intensity of the constraint. Carbon credits finance CDM projects. They are only paid *ex post*, once the emission reductions have been made, whereas allowances are defined *ex ante* in accordance with the emissions reduction targets imposed on the actors concerned. The articulation of these two mechanisms can function properly only under certain conditions: if there are too many projects and insufficient constraint by way of the emissions cap, the price of carbon will tend towards zero.

Linking a project mechanism to a cap-and-trade scheme has a threefold advantage. Firstly, at the environmental level, it is the average concentration of greenhouse gases in the atmosphere that affects the climate, and the geographical distribution of emissions makes no difference. Reducing emissions by a tonne of CO_2 from a village in Karnataka, from a port in South Korea, or from any other point on the globe brings exactly the same climate benefit. Linking up a cap-and-trade scheme such as the European Union emissions trading scheme (EU ETS) to a projects system therefore enables the pool of potential reductions to be enlarged.

Secondly, for actors subject to constraint, the system has the advantage of reducing their costs. The larger the pool of potential reductions, the greater is the probability of finding emission reductions at a low cost. And the disparities can be attractive: reducing methane emissions from a Chinese mine by equipping the tunnels with recovery systems amounts to around €2 per tonne CO_2 equivalent; recovering the methane given off by landfill sites in Brazil comes to around €5 per tonne CO_2 equivalent. These costs are considerably lower than what European companies have to pay in order to buy a CO_2 allowance on the market.

Thirdly, a project-based mechanism serves as a way of learning about and disseminating low-carbon production methods. It enables actors who are not constrained by regulation to exercise their innovative and imaginative capacities to find new, economically advantageous sources of emissions reduction. It also helps spread these methods beyond the economy operating under a carbon constraint.

The first carbon offset operations preceded the introduction of regulations by EU member states. They were launched twenty years before the start of the Kyoto protocol by companies, generally high CO_2 emitters, which planted new forests in order to reduce their carbon footprint.[1] This type of voluntary offset by directly setting up projects is not easy to undertake: one does not become a forest manager overnight when one's basic business consists of manufacturing semi-conductors. A small voluntary offset market was thus created enabling specialist operators to develop reduction or carbon sequestration projects by selling emissions reduction certificates. It is by using this type of mechanism that forms of voluntary offsets are today offered to the general public. Most car rental agencies or airlines, for example, provide simple ways on their websites to offset the emissions linked to journeys. An online calculator shows how much CO_2 is associated with the intended route. It is then simply a matter of giving one's credit card number in order to purchase the equivalent amount of carbon credits. These credits have previously served to finance a project reducing greenhouse gas emissions somewhere in the world.

---

[1] The majority of these companies have been American. In Europe, the Franco-Italian electronics components manufacturer ST-Microelectronics was one of the first to become involved in voluntary offset.

Despite advertising 'carbon neutrality' which they purportedly achieve, such voluntary offset mechanisms play a limited role in the carbon economy, with the exception of forestry projects. The genuine impetus given to project-based mechanisms occurred as a result of the rules established by the Kyoto Protocol.

The rules of the Clean Development Mechanism

Bagepalli and Osan are part of the first round of projects launched through the CDM. The CDM, introduced by Article 12 of the Kyoto Protocol, aims at using the carbon price to encourage developing countries that have not committed themselves to capping their emissions to accommodate or launch emissions reduction projects on their national territory. These projects may, subject to certain conditions, be awarded carbon credits. What is the purpose of these credits? The Kyoto Protocol envisages their being used to ensure compliance by countries that have committed themselves to capping their carbon emissions. In practice, the purchasers of carbon credits are not countries, but European industrial companies subject to the cap-and-trade system as regards their emissions.

To acquire their carbon credits, the holders of the Bagepalli and Osan projects have first to obtain the consent of the host country. This consent can only be given after public consultation, and subject to the project characteristics conforming to the host nation's development strategy. Both the Bagepalli and Osan projects met the development criteria and the projects were agreed by the Indian and South Korean authorities.

Once this stage has been completed, the project must be registered with the United Nations (UN) agency in Bonn responsible for the mechanism.[2] Registration takes place after a process in which the

---

[2] As we shall see in more detail in Chapter 8, the agency in question is the Secretariat of the United Nations Framework Convention on Climate Change

project holder must calculate the expected emissions reduction and get it validated by a third-party auditor accredited by the UN. The calculation is based on a comparison of the expected emissions once the project is implemented with those that would be produced in the absence of the project under a business-as-usual scenario. In the case of Bagepalli, the main problem was estimating the greenhouse gas emissions when very little is known about the villagers' energy consumption. In the case of Osan, the calculation was child's play for engineers professionally experienced in this type of exercise. The project developer must also demonstrate the 'additionality' of the project, i.e. prove that the existence of the carbon price makes its implementation possible. The additionality of Bagepalli is hardly debatable. Osan's additionality was justified by looking at the standard practices of similar types of factory in Asia, which do not eliminate nitrous oxide under the present economic conditions.

Registering the project with the UN is a crucial step for the project holder, for until such time as this is done it is totally impossible to obtain any credits. Nevertheless, the bulk of what remains to be done follows after registration. Credits are delivered only when emissions reductions have actually occurred, as a result of the implementation of the project. To obtain them, the project holder must submit a monitoring report, that is examined by another accredited auditor who attests that the reductions have taken place. The credits are then transmitted electronically to the project holders' accounts, which have been opened on an electronic register. The obstacle course is now complete. The carbon credit has been produced.

The UN is sometimes criticized for having set up a complicated, somewhat labyrinthine system to launch the CDM. Admittedly, there is still progress to be made, especially in terms of simplification and

(UNFCCC), which was signed in December 1992 in Rio de Janeiro and came into force in March 1994 after its ratification by the requisite number of signatory countries.

standardization. But the criticism is disproportionate. If one wants to ensure the environmental integrity of a project-based mechanism and not hand out credits indiscriminately, then it is essential to set up a system having eligibility rules and requiring project verification. The CDM has done this through a set of rules that establishes an internationally recognized standard. If it is at times over-sophisticated, this is a blessing for the various intermediary specialists, such as consultants, auditors, bankers, etc. But these rules have enabled the international carbon credits market to be started up.

The impact of large industrial projects

The first CDM projects were launched soon after the turn of the century. As of early 2010, 2,000 projects had been registered with the UN and 3,000 were pending. On paper, these projects should be able to yield emissions reductions of around 2.8 million tonnes CO_2 equivalent by 2012. This figure amounts to 6 per cent of the world's 2005 greenhouse gas emissions (including deforestation), or three years' growth in global emissions. In view of possible setbacks and delays, it is more reasonable to count on 1.2 billion carbon credits being delivered by 2012.

A little more than a quarter of these emissions reductions will be made by a handful of large-scale industrial projects, each accounting for several million tonnes CO_2 equivalent a year. They mainly concern chemical industry projects such as Osan or those reducing industrial hydrofluorocarbon (HFC)-type fluorinated gases in the refrigerants industry. These industrial projects have limited impact on the economy of the host countries and none at all on their energy system. But carbon credits enable them to get easy returns and sometimes amount to real windfall profits. It is for this reason that these projects accounted for up to 75 per cent of the first credits generated. The proportion is now rapidly declining and their development potential is limited.

Methane recovery projects in landfill sites and mines were among the first to be developed. They have been less successful than expected, but could account for 15 per cent of emissions reductions by 2012.

Projects developing renewable energy are likely to contribute around 35 per cent of the expected emissions reductions. Except for some hydropower projects, they are relatively small-scale projects, but cumulatively they can generate significant volumes. In this way, the project-based mechanism has an effect on the energy supply of the host countries. Projects intended to improve energy efficiency have taken longer to get going. They have considerable potential, but are held back by methodological problems.

Very few projects for reducing emissions in the transport sector have been registered. Apart from the difficulty of measuring emissions, the development of projects in this sector is hampered by the high price of CO_2 required to secure an adequate return on investments.

The CDM can, however, be used to help finance modal transfer operations in large cities, when investment entailed is not too large. The first such project registered involved creating a dedicated track system for bus transport in Bogota.

Lastly, attention should be drawn to the low level of development of projects in agriculture and forestry, despite their importance for greenhouse gas emissions in developing countries. As regards agriculture, a mere hundred or so projects were launched in the first wave of CDM projects. These mainly concerned the processing of livestock excrement in intensive rearing conditions, enabling methane to be recovered – biogas projects such as the one at Bagepalli producing far fewer credits – and recovering energy from certain crop by-products. There have, however, been no projects linked to changes in cultural practices or to atmospheric carbon storage.

The situation in regard to forests is no better. The rules developed in the context of the CDM to enhance their carbon storage capacity applies only to tree plantations. To manage the risk of the possible reversibility of storage, they are based on an extraordinarily complex system of temporary credits that buyers have shown little interest in. Despite the support of the World Bank, which has launched a fund dedicated to this type of project, only about fifty forest projects were at the validation stage at the start of 2010 and none of them has yet generated any credits.

Over the past few years, carbon credits have resulted in projects being launched that represent a sizable volume of emissions reductions in developing countries. This success testifies to the responsiveness of the economy to the appearance of a price. These projects started in the area where reductions are most profitable, viz. industrial gas and methane recovery. They still have had only a limited impact on the energy system and a marginal impact on agroforestry.

China, India and other countries

Where are the producers of carbon credits to be found? As of early 2010, some seventy countries had registered at least one project with the Secretariat of the UN Framework Convention on Climate Change (UNFCCC). But this figure is deceptive. In reality, production of carbon credits has been initiated by a limited number of emerging countries. It is anticipated that just four countries – China, India, Brazil and South Korea – will provide 80 per cent of the credits coming onto the market between now and 2012.

China is poised to produce more than half the credits during this period, its share of transactions having reached no less than 70 per cent in 2008. But Chinese domination of the carbon credit export market is nothing in comparison to its domination of the toy or textile markets.

The size and general economic attractiveness of the country has of course contributed to this state of affairs. But China has been very proactive in generating this production of credits. It has set up a dedicated administrative authority, with substantial resources and above all a stable set of rules for selecting projects. These rules have certainly not been designed to make entry into the market easy. For example, it is impossible for a foreign operator to implement the CDM in China without going through a company with a majority Chinese stake in the project. Moreover, China is the only country in the world to have introduced a tax on carbon credits produced on its territory. The base rate is 2 per cent for each new credit, but it rises to 65 per cent for projects incinerating HFCs emitted by factories. China is the leading world supplier in this respect, with eleven HFC-incinerating projects expected to produce 400 million credits between now and 2012. The revenue for Chinese public finances is likely to be close to €2 billion, entirely reallocated to the 'greening' of the economy.

Over and above these industrial gas reduction projects, China is rapidly enlarging its project portfolio. Investment in renewable energy is expected to amount to a third of Chinese production of carbon credits by 2012. It has given the go-ahead to more than 600 hydropower projects and more than 200 wind energy projects. In recent years progress has been made in terms of energy efficiency and fuel switching. And in 2006, with aid from the World Bank, China succeeded in registering the first carbon sequestration project through reforestation.

All in all, China's carbon credits could amount to export revenue of around €15 billion by 2012. On the assumption that the price of carbon represents a quarter of a project's cost, the CDM should give rise to emissions reduction investment of around $60 billion by 2012. Even on the Chinese scale, this is not insignificant.

India is expected to account for 16 per cent of the emissions reduced through the CDM. Its project portfolio is more diverse in quality than China's. In addition to industrial gas emissions, it covers many small projects involving bio-energy, such as the Bagepalli biogas project, and a large number of wind farm projects. In all, the sale of its credits up to 2012 could raise around €3 billion if they are put onto the market correctly.

Brazil's share of CDM credits up to 2012 is likely to be 6 per cent. It has a strong portfolio of projects in hydropower generation and landfill gas reduction as well as several industrial gas projects, but nothing in regard to forests. South Korea, with just seven projects registered, will be the fourth largest carbon credits producer during this period, with 5 per cent of the market.

In total, the first generation of projects should result in substantial financial transfers by 2012. The World Bank has set up a transactions tracking system for Kyoto credits. According to its estimates, the selling of credits over the period 2002–7 generated €44 billion in investment, of which €39 billion were for clean energy projects alone. Over the whole of the period up until 2012, the sale of carbon credits could generate €20 billion in direct currency inflows to the emerging countries and result in investment of some €60 billion to €80 billion.

The lessons of Joint Implementation

The Kyoto Protocol gave birth to a second project-based mechanism, the first credits from which began to be issued in 2008. Joint Implementation (JI), as it is known, defined in Article 6 of the Protocol, allows credits to be given for emissions reductions obtained by projects implemented on the home territory of developed countries or of those in transition to a market economy (Russia, Ukraine and eastern European countries) which have made a commitment to reduce their emissions.

JI was initially conceived as a way of encouraging emissions reduction projects in countries in transition to a market economy. The underlying thinking was similar to that of the CDM: these countries can deliver credits for projects launched on their territory that have demonstrated additionality of the emissions reductions obtained. Once issued, JI credits can have the same potential uses as credits resulting from the CDM. They are thus fungible with the latter.

While the basic principles are the same as those regulating the CDM, certain JI rules are specific to it. In particular, there are two possible ways of registering a project: one goes through the UN agency responsible for the system; the other arises from direct agreement between two countries that have taken on these commitments. In addition, JI credits can only be issued over the 2008–12 period covered by the Kyoto Protocol commitment. They do not apply to periods as long as those of the CDM, and this acts as a brake on their development.

As of early 2010, there were an estimated 340 million tonnes worth of JI credits exportable from Russia and Ukraine. The sums are modest in comparison to these two countries' known reduction potential up to 2012. Most of the projects are concentrated in the energy sector. Half of Russia's JI credits are concerned with reducing leakage from natural gas production and pipeline installations. There are also a significant number of projects aimed at substituting wood or gas for coal in heating systems.

In general, JI has a much lower profile than the CDM. Yet this family of projects offers lessons for the future. It shows firstly that project mechanisms can be used to reduce emissions in countries that have committed themselves to capping their emissions. Such a commitment is still only theoretical for Russia and Ukraine, whose authorized emissions limits under the Kyoto Protocol are very generous. But this is not the case for other countries that have used JI on their territory, such as Germany, France, New Zealand and Sweden.

Some JI projects implemented in the EU show that project mechanisms can become a good way of encouraging local municipalities or regional actors to participate in emissions reduction. Experiments of this kind have been launched in Germany and France, both of which in 2006 incorporated the JI rules as decreed by the UN into their national legislation. Some interesting innovations have occurred as a result, in particular the so-called 'programmatic' methods, which enable smaller-scale actors to access the carbon price system.

North Rhine-Westphalia is Germany's most densely populated *Land*. It is home to the largest area of heavy industry in Europe: chemicals, steel, engineering, coal, car manufacture, etc. Large installations have been encouraged to reduce their emissions since the launch of the EU emissions trading scheme in 2005. But the giants of German industry work with countless small-scale suppliers and sub-contractors that are not in the system and are not encouraged to reduce their emissions. It is this situation that led *NRW Energie Agentur*, the regional agency for energy efficiency, to set up a JI project targeting these small companies.

The agency provides a number of standard techniques designed to improve the energy efficiency of installations, with or without fuel substitution. These methods allow relatively simple emissions reduction calculations to be made. The actors adopting them can benefit from carbon credits. As well as delivering credits, the agency manages the programme and in particular is responsible for aggregating the basic operations and verifying emissions reductions. The aim in the first stage is to cut emissions by 60,000 tonnes of CO_2 a year. If the experiment is successful, such programmes could have a much greater impact in future.

Around ten such programmatic projects have so far been registered in Germany. Some of them cover small companies and others provide incentives for energy efficiency in private households. Up until 2012 the emissions reduction arising from these projects will remain relatively low. But their potential effects are considerable, especially if the lessons drawn from them are applied to the CDM. Consider the increasing part played by urbanization in the growth of emissions in countries such as China and India. Much is at stake in finding project-based mechanisms that award credits to cities making emissions-reducing planning and transportation choices. Let us imagine now that the Indian state of Karnataka had a regional agency like that of North Rhine-Westphalia, and used such credits to finance a biogas recovery programme in villages. The impact of Bagepalli-type projects would be greatly magnified. But steering the CDM in this direction would first require adopting a programmatic approach and would in particular mean modifying one of the key rules for registering a project, namely additionality.

How can project-based mechanisms be improved?

Setting up Kyoto projects under the aegis of the UN has in just a few years changed the scale of carbon offsets by establishing a global standard. This standard has led to the creation of funds investing in carbon credits and has been recognized by the benchmark market, the EU emissions trading scheme.

The standard's basic rule is that each holder must demonstrate the additionality of his project. The idea is to avoid windfall profits by issuing credits to actors who have no need for them. In general the additionality of a project can be shown by calculating its viability with and without a carbon value. In practice, any experienced consultant will put the right figures into the right boxes to get the desired result: carbon credits enable a project that is at the outset insufficiently viable to be hoisted up to the threshold of commercial profitability. The important thing is that the project does not appear sufficiently viable initially.[3] The incentive is not really up to the job!

The difficulty of finding criteria that are not arbitrary complicates the work of the UN agencies responsible for issuing credits and makes accredited consultants and verifiers happy by enabling them to invoice their services at generous rates. This in turn raises administrative costs and is an entry barrier for small projects, despite the existence of simplified procedures that are reserved for them. A fairly simple way of overcoming this contradiction is to shift the level at

---

[3] Another method is the so-called 'entry barrier' approach. If the investment you provide is not implemented in the present economic and social conditions, there is an entry barrier. If your credits enable it to be lifted, then you are additional. Here too, the arbitrary element is still considerable.

which additionality is set: instead of estimating it project by project, it could be established at the level of programmes bringing together basic projects or at a level of sectoral policies. In addition, such a shift could eliminate the unreasonable incentives criticized by certain economists.

As Michael Wara and David Victor of Stanford University have pointed out, it is easy to demonstrate the additionality of industrial projects such as Osan, which eliminate large amounts of CO_2 equivalent: the carbon price constitutes their sole revenue (Wara and Victor, 2008). Inherently in deficit without the price of carbon, they can become highly profitable through selling credits. As a result, the allocation of carbon credits risks encouraging investment in high carbon footprint processes in order then to gain credits by engaging in emissions reduction actions. This risk arose when the first carbon credits issued in exchange for HFC emissions reductions led to unscrupulous companies increasing capacity to collect carbon credits. However, such operators did not manage to obtain them because the project registration rules were in the meantime tightened up by the UN. Though effective in the short term, this solution is not appropriate in the medium term.

In this scenario, project-based mechanisms have a priming function. Once this function has revealed the sources of emissions reduction and their cost, project-based mechanisms should give way to other instruments. Let us suppose that Korea decides to build a second adipic acid factory. Now that the nitrous oxide incineration technology has been transferred thanks to the CDM, it should become the standard for new investments. The Korean government has to impose it and make companies which would not want to be in compliance pay for carbon. An interesting possibility, for example, would be to bring this type of installation into a European-style cap-and-trade scheme. Most industrial projects and almost all those involving energy production fall within this first scenario: project-based mechanisms enable the initial incentives to be given, but should then give way to more ambitious instruments within the framework of sectoral approaches.

Yet another risk arising from the present rules is that of taking away the host country's responsibility by not giving it an incentive to commit itself in the face of climate change. This risk led the French economist Jean Tirole (2009) to describe the CDM as 'a mistaken good idea'. In the current system, if a project involves investing to put an installation in compliance with existing regulations, it cannot count as additional. Take the example of landfill gas management. If the country hosting the projects introduces regulations obliging all landfills to recover this gas, the corresponding project will not be eligible for Kyoto projects mechanisms. They are no longer additional since, in the absence of carbon credits, managers must in any case be in compliance with the regulations in force. If the host country wants to retain the export revenues from carbon credits, it is in its interest to defer any regulation in this area, something that is counter-productive from an environmental angle. This type of incentive is rightly criticized by environmental organizations. It is one of the reasons for their antagonism to project-based mechanisms.

The 'programmatic' approach could, on the other hand, restore faith in public policies by shifting the level at which additionality is reckoned. Under the present rules, it has to be shown that a Kyoto project is not already in compliance with existing regulations. With a programmatic approach, it would be a matter of deciding whether the local government is able to set up regulations leading to measurable and duly verifiable emissions reductions. If so, the local government would then receive the carbon credits and would distribute them to actors who had reduced their emissions. Naturally, doing this requires an effective management system, with specified objectives and reliable measurement of results. In particular, a reference scenario would need to be defined and validated, in relation to which emissions reductions and the amount of credits allocated would be calculated. Various approaches of this type could be implemented after 2012.

In Europe, a programmatic approach should lead to the financing with emissions allowances of those industrial actors who first demonstrate carbon capture and storage along the lines of what will be introduced at Belchatow. At an international level, discussion of Reducing Emissions from Deforestation and Forest Degradation (REDD) mechanisms lies directly within this perspective.

The return to agricultural and forestry projects

The agroforestry system was not included in the first wave of carbon offset projects. Its exclusion is the major weakness of the projects mechanisms defined by the Kyoto Protocol. To correct this situation, it is important to make a distinction between the case of agricultural emissions reduction and that of carbon storing and release in the soil and forests.

Agricultural emissions reduction projects are eligible for the Kyoto Protocol's flexible mechanisms, and once carbon credits are issued they can be used for compliance by European companies on the carbon emissions trading scheme. The only problem is that there are very few projects, because the rules for registering projects are particularly unsuited to the agricultural sector. Moving to programmatic approaches is likely to facilitate the emergence of projects, as long as ways can be found of drastically reducing the administrative costs. Another condition is that the emissions avoided be calculated on a statistical basis and not for each separate economic unit as in the case for energy. The programmes concerned would need, therefore, to aggregate enough projects to have a robust sample in statistical terms.

The case of carbon sink projects is different. Here the rules defined by the Kyoto Protocol are of Byzantine complexity, which makes them almost impossible to apply in practice. Only carbon storage from a change in soil use by planting trees can be credited. But to counter the risk of storage reversibility, only temporary credits are issued. These credits are not accepted on the European market, which for its part has not drawn up a coherent set of rules enabling the protection of carbon sinks to be heightened.

When a carbon sink is in equilibrium, the flows of imported and exported carbon are equal. Carbon exports can be recovered for energy use, for example by the selective logging of old trees. Here one is producing bio-energy, which is renewable. This biomass can increase the return on carbon sinks all the more, since there is an increase in value of energy-related CO_2 on the emissions trading market.

During carbon sink formation, areas with growing trees and agricultural land that is being enriched with organic matter remove CO_2 from the atmosphere. These withdrawals offset the effects of emissions arising from the combustion of fossil fuels. Every additional tonne of CO_2 stored in forests or soil is equal therefore to an emissions reduction of a tonne of CO_2 of fossil origin. It would make good sense to set up a credit system for these inputs, which are significant when a forest is young and well-managed. But in order to be able to credit the replenishment of carbon sinks, there needs to be a symmetrical mechanism ensuring that in the case of capital destruction, the value of the sink is refunded by the payment of the corresponding credits. In an economy subject to carbon constraints, an economic player who destroys a carbon sink must pay for the value of the capital that goes up in smoke.

In the EU emissions trading scheme, very straightforward conventions for carbon sinks have been adopted. Since all European forests are young and growing, it is deemed that the equilibrium of carbon sinks is not threatened by using biomass for energy production. The use of wood, and more generally biomass, is therefore in all instances treated as non-CO_2-emitting. There is no case-by-case verification that the carbon sinks from which this bio-energy has been extracted are being renewed. This does not matter much so long as there are no imports from outside of Europe. But with the possible importing of biofuel, this becomes a real problem. The underlying reason is that the economic instruments set up in Europe have not included rules adapted to the valorization and protection of carbon sinks.

Logically enough, this limits the possibilities of using the carbon price to encourage good forestry practices. In the absence of such a mechanism, the European emissions trading scheme can therefore not deliver credits for carbon storage by the sinks. The experiments in this area conducted by forestry actors are therefore targeted in voluntary markets in which the risk of reversibility is generally hedged by the payment of premiums into guarantee funds.

Consequently a major issue in international discussions on REDD is how to find post-Kyoto rules that correctly take into account carbon storage-release mechanisms in the soil and trees so that the corresponding credits may acquire a real value. Such value could result from a commitment by public investment funds similar to those set up under the impetus of the Norwegian government or through the admission of these credits into the carbon compliance markets.

Project-based mechanisms and carbon markets

Carbon credits only acquire economic value if they find a buyer willing to pay a certain price for using them. The carbon credits market was started early in the first decade of the century by the World Bank and the Dutch government. With its Prototype Carbon Fund, the World Bank brought together a number of pioneer investors, countries subject to Kyoto constraints, and European and Japanese energy companies, all wanting to build up a carbon credits portfolio. The Dutch government set up a purchasing mechanism on its own account, operating according to the same principle of putting in a bid for credits very early in the process, at a time when prices were low and the risks high.

The market began taking off from 2005, following the launch of the European carbon trading system and the coming into force of the Kyoto Protocol. A secondary market where project initiators and companies subject to emission caps trade credits developed in Europe. The price of credits is correlated with that of the CO_2 allowances traded between industrial companies. It is nevertheless subject to a discount compared to the price of allowances, due to the uncertainties introduced by the existence of a limit on the use of credits by industry. But the basic economic mechanism is that the price of European CO_2 allowances has carried along the price of credits worldwide. The emissions reductions generated by the Bagepalli methanizers or the nitrous oxide incinerator at the Osan plant can be converted into credits that have a market value, because electric power companies thousands of kilometres away such as Drax and Belchatow are now required to pay a price for emitting CO_2.

The European market, and the rules of the Kyoto Protocol, provide visibility to sellers of carbon credits (excluding sequestration credits pertaining to forests) until 2012. After 2012, the rules have yet to be established.

The global scenario into which the Kyoto Protocol project-based mechanisms will be integrated will depend partly on how international climate negotiations evolve. It will also depend on the rules that will be set up in the various domestic carbon markets, especially the US market. One of the features common to the various working drafts so far discussed in Congress is the position given to project-based mechanisms, with explicit priority for agricultural and forestry projects. Depending on whether or not these rules turn out to be compatible with those of the system inherited from Kyoto, the impetus given to projects will be very different.

One of the interesting features of project mechanisms is that they can act as a link between markets operating with different internal rules. JI and CDM credits can thus be used both by governments and for the compliance of European and Japanese companies operating in very different institutional environments. This bridging role helps keep markets fluid and tends to reduce the price spreads that can arise because of different constraints from one market to the next. It can facilitate the transition towards a more integrated system in which the allowances stemming from emission caps become directly fungible.

An indispensable complement

In the carbon economy, offset projects play a complementary role. Their development does not directly make the objectives aimed at more ambitious, since these are determined when emissions caps are set. But it enables the potential scope of emissions reductions to be widened by bringing carbon value to actors who are not subject to caps. This role requires that strict measurement and verification rules for emissions reductions be established. If this is not done, the use of credits under offsets on the compliance markets is transformed into straightforward carbon leakage.

With its two project mechanisms, the Kyoto Protocol changed the scale of carbon offsets by creating an international standard. This standard allows industrial companies and governments to lower the cost of compliance by financing investments that reduce emissions, mostly in developing countries. It is based on a wealth of unmatched expertise and experience and on the trust inspired by its rules for issuing credits.

Because of its complexity and its cost of implementation, in its initial stage this standard favoured industrial projects. It has not had any ratchet effect on the agroforestry sector, which until now has remained beyond its reach. But the standard will evolve. For the industrial and energy sectors where it plays a priming role, it is likely to give way to more constraining instruments. Through new-generation programmatic-style projects, it could play a more lasting role in non-localized emissions sectors such as agriculture, forestry, transport and buildings. The programmatic principle may lead to the issuing of credits that remunerate genuine sectoral policies reducing emissions in a measurable and controllable way.

The REDD mechanisms designed to credit avoided deforestation could be the first prototypes of these. Their effectiveness in practice depends on their capacity to act on the causes of deforestation, which are very much related to agriculture. Deploying them on a large scale requires that they are not converted into subsidies that exempt from all financial responsibility the actors who are destroying existing carbon sinks.

The projects system inherited from the Kyoto Protocol may come to play a binding role, helping make the various regional carbon markets more fluid. It may in the future form an antidote to the risk of the fragmentation of markets resulting from insufficient coordination of climate policies. It has therefore very much a place in the market in regard to more extensive and effective pricing of greenhouse gas emissions worldwide.

SEVEN

Macroeconomic impacts:
sharing carbon rent

The town of Saumur on the river Loire is best known for its internationally reputed horse-training school, the Cadre Noir. Opulent residences rise up in tiers between the gently flowing river and the château perched on the hilltop. Time here seems to stand still. One can easily imagine bumping into characters from Balzac on the street corner. In fact the town is the setting for one of his most famous novels, *Eugénie Grandet*. Old father Grandet, the town's former cooper and a central figure in the novel, is the prototypal rentier or person of private means.

During his youth Grandet contributed to local prosperity through his barrel-making business. With the passage of time, he becomes a rentier. His properties are among the best located in the area and are particularly well served by public highways: a source of economic rent. Grandet's wine storage capacity enables him to sell his stocks when the vats of other producers are empty: a source of scarcity rent. His meadows bordering the Loire are more fertile than other land thanks to the alluvial deposits carried down by the river. Grandet therefore cut down the poplars that were covering the pasture in order to obtain this additional income: differential rent. When he is not contemplating his gold at nightfall, Grandet counts and re-counts his debt securities, which bring him fixed returns. Indeed the character has become ill from his rents. Because of his greed, he can neither benefit from it himself nor have his daughter Eugénie benefit from it. She is one of the most pitiful characters of *La Comédie Humaine*.

In literature, as in everyday life, rent and rentiers do not have a good reputation. Rent is associated with the idea of unwarranted gain or acquisition without effort. In economics, the concept of rent does not have this pejorative connotation. Its workings were revealed by the fathers of political economy, Adam Smith and David Ricardo. Without them, we would not be able to understand the economics of raw materials. Carbon is no exception. In Europe the launch of the carbon market led to a new economic value appearing: 2 billion tonnes of carbon dioxide (CO_2), with no value before 1 January 2005, has since then acquired a value ranging from €20 billion to €50 billion depending on allowance prices on the market. Where does this new value created out of nothing come from? The answer is that it is a scarcity rent: carbon rent. To properly grasp the impacts of the introduction of a carbon price into an economy, we must understand how this rent is divided up among the various players.

What is carbon rent?

Rent is associated firstly with scarcity. Let us take the case of oil rent. Oil is a non-renewable commodity. It is therefore characterized by a certain scarcity that increases with time as oilfields are depleted. This scarcity is expressed by a rent that is added to other elements of the cost of production (including return on capital) during price formation. Scarcity rent may be increased in the short term if there is an understanding among the main producers to restrict supply. In the long term, it increases at the rate of decline in the reserves left underground.[1] The same scarcity mechanism is the basis for property rent: a square metre is worth €50,000 on the Champs-Élysées, €1,000 in Aurillac in the mountains of central France and nothing in the Sahara, because space is scarce in the heart of Paris and abundant in the desert.

With the emissions trading scheme, Europe simultaneously created a price for CO_2 and a new scarcity rent amounting to some tens of billions of euros. This carbon rent adds to the production costs of CO_2 emitters.[2] If Europe decides to increase its commitment by tightening the constraint on industry, the quantities subject to allowances diminish and the value of the rent increases. The global value of rent tends to increase if the change in the price is more pronounced than that of the quantities, which is generally the case.[3] Vice versa, if Europe relaxes its rules, the rent decreases and can even disappear if the price of carbon falls to zero.

A second major rent mechanism is linked to the difference between the costs of production. This aspect was brought to light by Ricardo on the basis of differences in land fertility, similarly to that which enabled Grandet to profit by cutting down the poplars bordering the Loire. It is a matter here of differential rent. Differential rent enables us to understand the rules for distributing rent among the various producers. Let us illustrate this by considering oil production. The cost of extracting a barrel of oil from the Ghawar oil field in Saudi Arabia,

[1] The economist Harold Hotelling showed that to be optimal, the growth rate in the value of the rent has to converge towards the economy's long-term interest rate.

[2] This emerges clearly in the case of an allowance that is paid for. In the case of a free allowance, firms subject to quotas carry what economists call an 'opportunity cost'. The opportunity cost is the cost of forgoing a benefit. Once allowances are transferable and have a market value, a firm that decides to keep them or use them for compliance forgoes a potential revenue on the market.

[3] The marginal abatement cost is indeed increasing since the emissions reductions are first made where they are least expensive. To put it in more technical terms, the emissions reduction marginal cost curve is convex.

the world's largest, is around $2. If oil is trading at $100 a barrel, then the rent amounts to $98 per barrel extracted. In the North Sea, this cost is around $20, therefore the rent decreases to $80. In wells drilled in deeper water or to extract shale oil, considerably more has to be spent. The rent shrinks further and disappears altogether for deposits where extraction is most costly.

We again find this differential rent situation in the sharing of carbon rent among European companies. To produce 5,000 kilowatt hours (KWh) of electricity – the average annual consumption of a European household – 5 tonnes of CO_2 are emitted with a coal-fired power station, 2.5 tonnes with a gas-fired power station, and none with a hydropower dam. Let us suppose that CO_2 is priced at €20 a tonne. An electric power generating company that shifts from coal-based production to gas-based production obtains a differential rent of €50 to produce these 5,000 KWh. If it changes dramatically to a non-emitting source such as hydropower or nuclear power, the differential rent rises to €100. The search for differential rent is the motivation that prompts these companies to reduce emissions once carbon has a price. It is environmentally virtuous.

When an emissions trading scheme is set up, the size of the scarcity rent created by capping emissions is generally much larger than the size of the differential rent obtained by emissions reductions. In the European Union emissions trading scheme (EU ETS), scarcity rent appeared with the emerging price of carbon, amounting to some €40 billion a year in 2005 and 2006, before the price fell to zero. The differential rent obtained through emissions reductions was around one-fortieth of this size. Figure 7.1 represents the two rents in the form of rectangle A and triangle B. As the cap is reduced, there is a shift to the left part of the graph and the amount of differential rent obtained through emissions reductions increases. But when the market was launched, the actors subject to constraint had a major concern: being able to recover the largest proportion of scarcity rent by obtaining the best possible initial allocations.

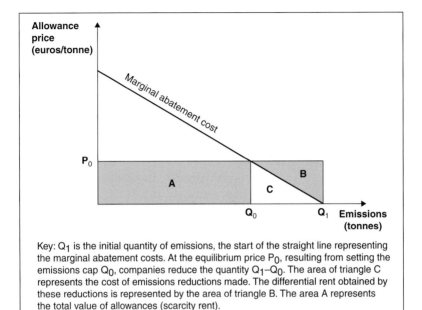

Key: Q_1 is the initial quantity of emissions, the start of the straight line representing the marginal abatement costs. At the equilibrium price P_0, resulting from setting the emissions cap Q_0, companies reduce the quantity Q_1-Q_0. The area of triangle C represents the cost of emissions reductions made. The differential rent obtained by these reductions is represented by the area of triangle B. The area A represents the total value of allowances (scarcity rent).

FIGURE 7.1: *Scarcity rent and differential rent in a carbon market*

Rent sharing on the European carbon market

The ending of a free good is rarely welcomed enthusiastically by those who have to pay the bill. The companies subject to allowances were quick to raise the spectre of a loss of competitiveness, or worse, relocation, with the introduction of a carbon price. In fact, the pricing of carbon has had some major economic consequences. But in no way do they amount to an additional charge that indiscriminately raises the costs of all companies.

During the market's trial period from 2005 to 2007, nearly all allowances were allocated free of charge. Every year, the companies concerned could freely use the carbon rent, embodied in the CO_2 allowances received, until compliance was achieved. Once they were in compliance, they had to pay for their emissions with allowances. In cases where they were short of allowances, the rent received was not sufficient to ensure the payment of all their emissions. In the reverse situation, they retained part of the carbon rent.

Outside the electricity sector, companies were net sellers of CO_2 allowances. They therefore kept part of the rent distributed with the allowances, amounting to some €2 billion. In addition to this first benefit, another benefit exists, resulting from the impact of carbon rent on selling prices. In industrial markets open to international competition, prices reflect international market conditions. It is therefore difficult to pass on the cost of carbon to the consumer. Given the tension of markets over the 2005–7 period, it is possible that some producers of major intermediate goods such as steel or construction materials were able to do so. In which case, they retained a further proportion of carbon rent by billing their customers for the price of freely allocated allowances.

Electric power companies were generally net buyers of allowances. The scarcity rent distributed in the form of free allocation of CO_2 allowances was not sufficient to ensure the return of allowances needed for compliance. The excess cost created by net emissions purchases by electric power companies was around €2 billion. But most of this amount was not paid for by the electricity industry, but by their customers.

Because it cannot be stored and is not easily transported, electricity is not subject to international competition in the same way as steel or cardboard. Historically, most electric power was sold at regulated prices set by the public authorities. This is still largely the case in France. But with the liberalization of the markets, given impetus by Brussels, a growing proportion of electric power is traded on markets where the equilibrium price tends to be set at the marginal cost, i.e. the cost of production of the last kilowatt hour put onto the grid. The provision of last kilowatt hours for the grid requires setting up thermal installations that are generally heavy emitters of CO_2. The marginal cost of electricity therefore has a high carbon component. In the prices billed to their customers, power companies add in part or all of the value of allowances that have been mostly distributed free of charge. Thus it is the customers who finally pay the scarcity rent initially distributed to the power companies in the form of free allowances.

During the first phase of the EU ETS, companies were the beneficiaries of the allowances system. Industrial companies subject to allowances benefited from a net transfer of financial resources from the public authorities, equivalent to a subsidy. This transfer was augmented by customers, mainly power sector customers not having had access to regulated prices or set prices in the framework of long-term contracts.

In shifting from a free system of allocation to a paying system, the public authority acquires the carbon rent when the allowances are sold, generally via auctioning. In the EU ETS, auctioning emissions will be the general rule for the electric power sector from 2013 onwards and is likely to be deployed more gradually in other sectors.

The first impact of moving to auctioning concerns the distribution of carbon rent within the electric power sector, where auctions could represent €20 billion to €30 billion a year between 2013 and 2020.[4] This scarcity rent will no longer be distributed to installations emitting CO_2, but retained by the public authority. The generating companies will still be in competition in the market in order to acquire the differential rents resulting from the provision of electricity from installations emitting no or very little CO_2. In France and Sweden, where 90 per cent of electric power is generated with no CO_2 emissions, a higher proportion of the carbon rent will remain in the hands of electricity companies. In Germany and the UK, where power generation is considerably more carbon-emitting, the state will take the largest share of the rent to the detriment of local electricity companies.

---

[4] In the appendix to Ellerman *et al.* (2010), Jan-Horst Keppler provides figures on the rents generated by pricing CO_2 in the electric power sector.

Secondly, the shift to the auctioning of allowances strengthens the incentive to invest in new, low-carbon equipment: the opening of new production capacity emitting CO_2 will be subordinated to the buying of allowances to pay for the corresponding emissions. This move corrects the incentive created by the national reserves system set up for 'new entrants' during the first phase of the market. These reserves make it possible for each EU member state to allocate free allowances to new installations, in other words to use the carbon rent to subsidize investment for building plants emitting greenhouse gases. The shift to auctions corrects this by encouraging low-carbon investment. The impact is beneficial for the economy if the volume of low-carbon investment thereby induced is greater than what would anyway have been generated by the need for renewal and extending the stock of installed production. If the reverse is the case, then the economic impact is negative.

The economic impact of the system will also depend on the use made of the revenue of auctions. Used wisely, auctions will correct the undesirable social impacts caused by pricing carbon, for example by providing income support for low-income households adversely affected by the higher price of energy from fossil fuels. The public authority can also use this new resource to correct current economic distortions, for instance by reducing employers' social security contributions when these represent an excessive proportion of labour costs or by helping to bring down an uncontrolled public deficit.

Once all allowances are sold by auction, the macroeconomic impact of the system is in fact estimated in the same way as for an environmental tax. Its impact is *a priori* neither positive nor negative for the economy. It all depends on how it is introduced and how it is used. The carbon price makes using energy derived from fossil fuel more expensive, which hinders the economy. At the same time it encourages new investment and enables the carbon rent to be reinjected into the economic circuit. The net effect on the economy depends on the modalities of this redistribution and on the behaviour of businesses and households.

Geographical distribution of carbon rent

Carbon rent does not only lead to the redistribution of wealth among actors. It can also be the source of substantial geographical transfers. The establishment of the EU ETS generated such transfers, both among European countries and between Europe and the rest of the world.

We saw this in a particularly dramatic way with the Greek financial bailout package in the spring of 2010: European solidarity is a generous and consensual concept. But once it becomes a matter of applying it and taking money from Peter and transferring it to Paul, the situation rapidly becomes conflictual. The creation of carbon rent through the emissions trading scheme nevertheless resulted in this type of shift of resources among European countries without giving rise to too much tension. The transfer operated through the market without appearing in national budgets or, ultimately, on people's tax forms. Because the allocation of allowances has on average been more generous in the twelve accession countries, these have been net sellers of allowances.[5] The resulting net inflow of currency stood at around €300 million a year on average between 2005 and 2009. This is not insignificant for these countries, where wealth per capita is less than half that of the more affluent countries of Western Europe.

Another major geographical transfer resulted from the opening of the European market to the Kyoto project-based mechanisms. In this case, European carbon rent is used as a lever to finance emissions-reduction projects outside the EU ETS. The great majority of offsets purchased have been CDM credits financing emission reductions in developing countries. Here again, the financial transfer goes to

---

[5] These are the ten eastern European countries together with Cyprus and Malta that joined the EU during the decade following 2000.

countries with the lowest per capita wealth where the cost of emissions reduction is lowest.

But this way of using rent should be carefully handled. If Kyoto credits are available without restriction, there is a risk of destabilizing the equilibrium between supply and demand on the market, thereby provoking a fall in the price of carbon. It is for this reason that the quantity of carbon offsets allowed to access the European market has been capped. In the period up to 2012, if this cap is reached, Europe could be transferring some €5 billion to €8 billion a year to developing countries and to eastern Europe (mainly Russia and Ukraine) by using part of the carbon rent to buy offsets on the international market.

Let us now change the scale and see whether what has been learnt from the analysis of carbon rent on a European scale can be applied globally.

Growth or degrowth?

One of the misgivings in regard to pricing carbon concerns the negative impact that it may possibly have on the economy. In certain cases, such concern is straightforward propaganda or defence of the short-term interests of high-emission companies. But apart from special lobbying, the idea that limiting the free use of the atmosphere imposes a cost that hampers the growth capacity of economies – especially those which are in most need of growth – is soundly based. It has led certain observers to conclude that action against climate risk and the development of poor countries are in competition with each other.[6] Is it really necessary to choose between long-term climate risk and short-term economic development?

---

[6] Bjorn Lomborg, for example, writes in his blog: 'Instead of saving one person from malaria through climate change policies, the same amount of money spent on malaria could save 36,000 people'. Richard Tol (2005) puts forward the same argument in a more academic form.

From a historical perspective, there is no doubt that the increase of emissions and economic growth are closely linked. This can be seen not only in long-term trends, but also in short-term economic fluctuations: greenhouse gas emissions grow when the economy is dynamic and fall during recessions. Let us look more closely at the case of recessions.

Apart from major wars, the 1929 crisis, which culminated in 1932, though some of its effects continued until the second world war, was the most violent shock to have hit the world economy in the twentieth century. According to Maddison's estimates, the Great Depression resulted in a 10 per cent decline of world gross domestic product (GDP) from 1929 to 1932, and a 27 per cent decline in the US, the true epicentre of the crisis. Global emissions of CO_2 resulting from energy production fell by 30 per cent over the same period. In the US they fell from 2 billion to 1.2 billion tonnes. The basic idea to keep in mind is that a recession tends to lead to disproportionate reductions in greenhouse gas emissions. This type of relationship between emissions and economic activity is found in all recessions that have marked economic history.

The two recessions associated with the oil shocks of 1973 and 1980 brought an end to the post-war boom, leading to falls in emissions of 2 per cent from 1973 to 1975 and of 6 per cent from 1979 to 1982. It is, however, difficult to separate what resulted from the recessions themselves and the effects caused by the increase in the oil price that preceded them. Another major twentieth century recession affected the countries that left the centrally planned economic system following the implosion of the Soviet Union. The fall in Russian production in the 1990s led to CO_2 emissions from energy production declining by a third, and in Ukraine by slightly more than half. Comparable declines occurred in eastern European countries following the fall of the Berlin Wall. More recently, the 2009 recession led to a reduction in emissions of around 10 per cent in the industrialized countries, considerably more than the decline in GDP.

This close correlation between economic growth and greenhouse gas emissions is grist to the mill of 'degrowth supporters'. Developed at a theoretical level by the Romanian-born economist Nicholas Georgescu-Roegen, the concept of degrowth is today defended by certain environmental movements. The idea is to cut back production of goods and services so as to reduce our ecological footprint. Some people also apply the concept to the population, advocating a systematic reduction in the birth rate, in a direct line of descent, from the classical economist and Anglican minister, Thomas Robert Malthus.

The degrowth option would consist in voluntarily and sustainably downsizing the economy in the interest of protecting the planet. Given the observations we have just made, this would unquestionably have a major impact on greenhouse gas emissions, which would fall sharply.

Many aspects of degrowth can be justified in wealthy Western societies. The quality of life is far from improving in line with the accumulation of goods and services, as has been demonstrated by the work of William Nordhaus and James Tobin in the US and Jean Gadrey in France.[7] The overabundance of goods can even give rise to negative effects, such as obesity, now a veritable public health pandemic. For these countries, degrowth above all means that we could improve collective well-being by adopting ways of life that reduce waste of all kinds and excess consumption. But degrowth hardly has the same meaning for all those people – and there are many of them in the world – who do not have access to basic goods and services: food, energy, mobility, health, primary education.

To reconcile degrowth and equity, it would therefore be necessary to limit the reduction in consumption to the rich. This approach is not very realistic: we should not count on the affluent giving up their

[7] The highly publicized report of the Sen/Stitglitz Commission ordered by President Sarkozy made a number of suggestions for taking account of these aspects of well-being in calculating indicators supplementary to GDP.

wealth to achieve greater fairness. And in any case quickly redistributing wealth in the name of equity is a far from easy task.

Another major difficulty would be to make degrowth compatible with the capacity for technical creation and innovation, one of the drivers of economic progress. To act effectively against climate risk, this creative capacity needs to be mobilized as much in the area of research and the spread of low-carbon technologies as in that of social innovation. The degrowth option does not seem appropriate. Rather we should try and reverse the correlation between economic growth and greenhouse gas emissions. Putting a price on carbon helps achieve this in two ways: it imposes an additional cost on greenhouse gas emitters, who now have a financial interest in reducing their emissions if they want to continue prospering; and at the same time it creates carbon rent, a new resource that can immediately boost economic development.

The egalitarian distribution model

Suppose that a carbon price of €20 per tonne CO_2 equivalent is applied to all the planet's emissions. Emitting 45 billion tonnes CO_2 equivalent into the atmosphere then costs €900 billion per year – the equivalent of nearly ten times the total public aid for development, or India's GDP.

One way of looking at this €900 billion is to see it as an additional cost weighing down the world economic system and hampering its growth. But there is an alternative, more compelling view, based on economic analysis, in which this €900 billion is not a burden on the world economy, but is instead a carbon rent that mankind has decided to create artificially to protect the capacity of the atmosphere to regulate the climate. As in this EU ETS, this rent moves resources between economic agents and its existence results in many distributive effects. Its impact on the world economy depends on the combination of two effects: an increase in the price of goods with a high carbon impact, which has a negative short-term effect; and the transfers resulting from the dividing up of the carbon rent, which acts as a stimulus to the economy. This second effect depends on the model chosen to distribute the rent when the global emissions cap is set. Let us start with a very simple model: strict equality, for every inhabitant of the planet, as regards the right to emit greenhouse gases.

In this egalitarian model, each country receives free emission allowances in proportion to its share of the world population. For example, the US, which accounts for 317 million of the world's 6.9 billion inhabitants, receives just 4 per cent of the total allowances, even though it emits a fifth of the world's greenhouse gases. It will therefore be a net purchaser on the world market. This model generates economic transfers from countries with high emissions levels per capita to those with low emissions levels. These transfers allow the rule of equality of emissions rights among the world's citizens to be respected.

We would end up with strictly the same distributive effects by setting up a world system of emission allowances auctions in which the product is redistributed to countries in proportion to their populations. Another way of operating would be to introduce a global tax on greenhouse gas emissions, the revenue from which would be redistributed on an egalitarian basis. Such a system is advocated by the American economist Steven E. Stoft.[8] He was inspired by the example of the oil tax in Alaska: each year in June, every resident receives a cheque from the state government in respect of the fees paid for using the pipelines. The value of the cheque is the same for each inhabitant of the state.

A €20 tax applying to all the world's greenhouse gas emissions would allow €130 to be given to each of the Earth's inhabitants. This sum is a trifle for an American manager, who will be required to spend

---

[8] Stoft (2008) recommends introducing a global tax on carbon, the revenue from which would be distributed equally among the world's citizens.

€470 to pay for his 23.5 tonnes CO_2 equivalent emitted over the year (plus the same amount for each member of his family). On the other hand, it is a significant amount for an Indian worker, who emits less than two tonnes CO_2 equivalent a year and therefore benefits from a net transfer of nearly €100 after payment of his carbon tax.

At a macroeconomic level, the redistributive effects of the egalitarian model are substantial. On the basis of figures for 2005, a €20 tax would result in net cash outflows of €93 billion a year from the US and net inflows of €110 billion a year to India. Globally, its introduction in 2005, the year when the Kyoto Protocol came into force, would have led to the transfer of €200 billion from the developed countries to the developing countries, considering the differences in the emissions level per inhabitant in the two groups of countries. In the egalitarian model, the introduction of a carbon price on a world scale would allow the resources transferred towards low-emitting developing countries to be substantially increased. This transfer is not deferred over time: it is concomitant with the appearance of carbon rent. Two economists at Columbia University, Graciela Chichilnisky and Geoffrey Heal, have shown that such a transfer to poor countries is a condition for a single carbon price to lead to an optimal situation in regard to welfare economics (Chichilnisky and Heal, 1998).

The probability of such a 'big bang' implementation, with no transitional period, of a shift from the present situation to an emissions cap egalitarian model is, however, infinitesimal, if not zero. Indeed the systems set up for pricing greenhouse gas emissions have followed a quite different model for the initial distribution of carbon rent: the so-called grandfathering model.

The grandfathering model

One of the reasons why the introduction of cap-and-trade schemes in the real world was relatively easy was precisely that it followed the grandfathering model rather than the egalitarian model. The grand-fathering model involves taking a historical reference point for emissions then choosing a target level in relation to this historical base. It was this method that was adopted in Europe for the ETS and, as we shall see in Chapter 8, the same system was used for the Kyoto Protocol commitments.

The distinctive feature of this method is that when the system is set up, it minimizes the redistributive effects. Indeed it leaves the major part of the carbon rent at the disposal of the largest historical greenhouse gas emitters. As well as facilitating their acceptance of the scheme, this gives them the financial means to convert their production system. Applied to countries, this method leads to most of the carbon rent remaining in those with the highest emissions rate per capita. The reason why it is so difficult extending the Kyoto accords to developing countries is that the grandfathering model is used to share out emission rights among developed countries.

To overcome this difficulty, certain people have put forward the idea of using a third measure to allocate emissions allowances on a world scale: GDP. A benchmark allocation model is then used, in which emission rights are distributed according to technical production coefficients. Applied to the steel or cement industry, the benchmark method leads to allowances being allocated in proportion to the installed production capacity while adopting the same CO_2 emissions ratio per unit produced for the whole of the industry. When applied to countries, the benchmark method shares out emissions rights by equalizing the carbon intensity of the different countries' GDPs. If, for example, the US emits twice as much CO_2 as Europe for a comparable GDP level, it obtains the same number of allowances and must cover its emissions by purchasing allowances on the international market. In practice, the use of GDP intensity coefficients in terms of greenhouse gas emissions to apportion carbon rent throughout the world would have undesirable distributive effects. It is unfavourable to developing countries, most of which emit more greenhouse gas per unit of GDP than the industrialized countries. Moreover, this rather unpromising system is also complicated and costly to implement.

The right way to extend carbon pricing while stimulating economic development is to combine the grandfathering model and the egalitarian model. The first allows pricing systems to be set up and tested, the rules of which can then evolve rapidly, as in the European case. The second is a long-term target which should be linked to medium-term commitments made on the basis of historical reference points. In the short term, this target seems out of reach and comes up against a great number of interests and established habits. A rigorous analysis of the mechanisms of carbon rent shows, however, that the transfer of resources through the egalitarian model can lead to both economic efficiency and equity.

Carbon rent and the protection of carbon sinks

In general, rent is associated with an asset and helps conserve that asset. For example, the ground rent paid to landowners enables them to maintain their properties. Should not carbon rent therefore remunerate the holders and users of carbon stocks in the ground or biosphere?

Carbon rent was created to protect a common asset – the atmosphere – against the accumulation of anthropogenic emissions. The atmosphere is the destination reservoir of human greenhouse gas emissions, and we must cease filling up this reservoir if we are to avert climate risk. What generates rent is the greenhouse gas emissions into the atmosphere. Its distribution among economic actors depends on choices made in regard to the allocation of allowances between the various emissions sources. This rent is largely independent of the distribution of carbon stocks in the ground or in the biosphere. Let us now look at these stocks from which carbon is drawn before being transformed into greenhouse gases.

The largest of these – the oceans – is, like the atmosphere, a global public good. But other reservoirs are owned and used either by private concerns or by governments, which control the resources in their territory and have no intention of sharing them. On the one hand, these concern agricultural and forest carbon sinks and, on the other, oilfields, natural gas deposits and coal mines.

The use of carbon rent to replenish agricultural and forest carbon sinks take us back to the discussion about the international mechanism that enables carbon credits to be issued to avoid deforestation. As we saw in Chapters 5 and 6, setting up programmes in the context of Reducing Emissions from Deforestation and Forest Degradation (REDD) can become a direct way of using carbon rent to finance the maintenance of carbon sinks in forests.

The question of hydrocarbon sinks has not been discussed to the same extent in the context of climate negotiations between countries holding carbon stocks and countries using fossil fuels to produce energy. Some local initiatives are nevertheless setting their sights on financing the formation or preservation of hydrocarbon sinks. The thinking is very much the same as that behind the European project for financing the first industrial demonstrations of carbon capture and storage by using the reserve built up from the sale of allowances to industry.

The oil deposits in the Ecuadorian forest provide another textbook example. Around 20 per cent of Ecuador's reserves are located in the Yasuni region, one of the ecologically richest areas of Amazonian forest and, as such, protected by a reserve. Nearly 900 million barrels of very heavy crude oil lie buried beneath the trees, equivalent to some 400 million tonnes of CO_2. Three wells might be drilled, which within a few years would be providing the Ecuadorian government with around €280 million a year for the concession. Under the impetus of President Raphaël Correa, the Ecuadorian government abandoned the project, but negotiated a novel contract with international sponsors. Under this contract half the anticipated revenues from extracting the oil will be paid to the Ecuadorian state in exchange for the avoidance of oil exploration.

For it to continue, and particularly for it to be extended to other situations, such payment for oil exploration avoided cannot be based solely on the volition of a political leader and the goodwill of foreign sponsors. It must be underpinned by sustainable mechanisms. Cannot the carbon stock beneath the ground become the asset backing the payment? At a market price of €20 per tonne, a stock of 400 million tonnes of CO_2 represents a capital value of €8 billion. A 3.5 per cent perpetual rent would indefinitely ensure a substitution revenue for the Ecuadorian state equal to the €280 million from the concession. In this case, carbon rent would replace oil rent for the greater good of the forest and the indigenous communities of Yasuni National Park.

Carbon rent versus oil rent?

According to calculations by Marie-Claire Aoun (2009) in *The New Energy Crisis*, global oil rent amounted to €1,600 billion in 2004. The consumer countries took a substantial proportion of this through fuel taxes, which is a pure form of monopoly rent. €670 billion was divided up among the international oil companies and €560 billion was retained by the producer countries.

In the short term, this oil rent is a highly coveted resource that has led to many a conflict as well as to the rapid enrichment of those who possess it. In the medium to long term, rent is the source of what has been called 'the resource curse'. Rentier economies can bring unexpected prosperity to their citizens, or at least to their rulers and middle classes, but they have trouble converting rent into investment that enables them to enjoy sustainable growth.

As we saw in Chapter 4, the extension of the price of carbon deriving from energy production will lead to a fall in oil prices. In other words, carbon rent gradually erodes oil rent and leads to its disappearance in the long term. This situation accounts for the reluctance of large hydrocarbon producers and exporting countries to commit themselves in regard to climate negotiations, for it is not in their interest to enter a constraining regime. How then can their participation be secured? The answer is, by distributing emissions allowances as far upstream as possible within the industry, i.e., where the oil or gas comes out of the ground.

Let us call this third allocation model the carbon sink model. In such a model, an allowance has to be surrendered for each tonne of energy-related CO_2 equivalent at the moment of its extraction. The global cap is set within the framework of a declining long-term trajectory commensurate with climate risk that has been negotiated with the holders of carbon sinks. It represents a certain proportion of the total stock of carbon in the ground. The distribution of allowances among the holders of these stocks is worked out in proportion to the carbon remaining in the stock. Every time an owner extracts a barrel, it surrenders the allowances corresponding to the CO_2 content of the oil and loses part of its allocation for the following year. The system therefore leads automatically to tightening the constraint on owners who extract carbon from beneath the ground while continuing to pay those who choose to let the oil remain in place.

One of the difficulties of putting this model into practice is the inadequate knowledge of existing carbon stocks, particularly in cases where new deposits are discovered. For allocation in accordance with the carbon sink model to work, a reserve needs to be created for 'new entrants', who every year have to possess allowances representing a certain proportion of their carbon stock. These allowances will most often be resold to the owners of deposits that are already being exploited. For example, every year Ecuador may sell a proportion of its carbon stock to Saudi Arabia. Thus the exploitation of the Ghawar oilfield will finance the protection of the Yasuni carbon sink.

Like the egalitarian model, the carbon sink model has little chance of being rapidly adopted on a large scale, if only because of the difficulty of obtaining the necessary information. It should nevertheless be discussed at negotiations, for it is one of the few levers that can attract the big hydrocarbon producers into cooperative interplay around the management of carbon rent.

The Gordian knot of international negotiations

The introduction of a constraint on greenhouse gas emissions is often seen as a new cost that would weigh on the economy, compromising its capacity for growth. As a result, decision-makers would have to arbitrate between long-term climate risk and medium-term development prospects. Yet rigorous economic analysis shows that one can largely escape this dilemma if one can manage to distribute carbon rent efficiently and fairly.

Pricing carbon stems from a human decision to artificially create a rent that reflects a new scarcity: that of the capacity of the atmosphere to absorb all human greenhouse gas emissions without risking the climate. The new carbon price indeed has a recessive effect on the economy by increasing the cost of using fossil fuels for energy production. But the scarcity rent created by this price is redistributed among the economic actors, which in turn encourages new expenditure, particularly as regards investment. The overall impact on the economy depends on the sum of these two opposed effects.

The emergence of a carbon price in Europe has not hampered economic growth. Companies subject to caps have benefited from net transfers from the public authority and consumers. Price increases attributable to the EU ETS were limited as a result of international competition and regulated electricity rates. The transfers generated by carbon rent have advantaged countries with lower income levels both within and outside Europe. The move to auctions and tightening the cap will increase the pressure on these companies and their customers. At the same time it will reinforce the incentives to invest in low-carbon technologies and heighten the role of the public authorities in the distribution of carbon rent.

At the international level, widening the pricing of carbon is likely to give rise to very substantial resource transfers, on which its global impact will depend. These transfers can pass through three channels: those arising from the initial allocation of emissions rights; those resulting from setting up project-based mechanisms; and those likely to be financed by public funds from auctioning.

The grandfathering allocation model, based on past rights, facilitates the acceptance of the system by large-scale emitters, but makes acceptance by emerging countries with the lowest income levels problematic. The allocation model based on equality of rights generates high and welcome resource transfers to developing countries, but its acceptance by industrialized countries is very unlikely. The model based on carbon sinks would allow hydrocarbon producers to be implicated in the management of carbon rent, but setting it up would be complex.

Carbon rent has immediate economic consequences that are much more crucial for political decision-makers than the effects of the carbon price on the level of future emissions. At climate conferences, where the future of the planet may hang in the balance, much time is spent haggling over how carbon rent will be shared out.

EIGHT

International climate negotiations

In his well-known paper 'The Tragedy of the Commons', Garrett Hardin (1968) describes the mechanisms through which natural resources are damaged as a result of being free of cost. He focuses on the example of the communal pastureland or 'commons' that surrounded English villages up until the end of the eighteenth century. Under this tenure system, the villagers all had access to communal grazing land for their livestock. In a situation of demographic stagnation, this social system provided them with a measure of security. Everyone had free access to a shared resource.

With a growing population, the system tended to destroy itself. Because access to the communal grazing land was free, in making their economic calculations the herdsmen took no account of the cost of this resource to the community. It was therefore in each herdsman's interest to have his animals graze on the common as long as a positive marginal revenue remained: a few blades of grass for the last animal taken to the pasture. Inevitably the outcome was overgrazing that reduced the fertility of the common to zero and led to the destruction of the collective good.

The complexity of international climate negotiations may be appreciated if one takes Hardin's example and replaces the word 'village' by 'planet' and the word 'common' by 'the atmosphere'. The growth of the planet's population and its rising living standards are a threat to a very special collective good: the stability of the climate. Like the village common, the atmosphere is not infinite. Its capacity to regulate temperatures is limited. Since the beginning of the industrial era, however, the growth of the economy has been based on free access to the atmosphere. Similarly to the villagers' use of the communal pastureland, economic actors continue to emit greenhouse gases into the atmosphere, thereby putting at risk the common good.

As long as the use of the common good remains free, it is in each actor's interest to continue using it as he sees fit, at the risk of leading to its destruction. If the villagers decide to organize themselves collectively in order to protect this good, it is then in each villager's interest to contribute less than his neighbour, since the cost of the action is individual and the benefit – the protection of the common – collective. Herein lies the problem facing the representatives of the world's 192 countries which meet each year at the United Nations (UN) annual climate conference. Is this 'mission impossible' in a world dominated by private interests and the selfishness of nation states? There is, however, a successful precedent as regards a collective approach to protect a collective planetary good, namely the approach taken in the framework of the 1987 Montreal Protocol to confront the risk created by the destruction of the ozone layer.

International collaboration to prevent the destruction of the ozone layer

Chlorofluorocarbons (CFCs) – halogenous molecules marketed under the trademark Freon – quickly established themselves in the post-war period in the refrigeration and air-conditioning industry and for aerosol sprays due to their stability, non-flammability and lack of toxicity. But these health and environmental benefits are local. During the 1970s a team of chemists, who in 1995 were awarded the Nobel prize, found that this type of gas liberated constituents that were liable to destroy the stratospheric ozone layer protecting the surface of the Earth from the sun's ultra-violet radiation.[1] It was the first alert from the scientific community directed at policy-makers. In 1985, a number of observations revealed the existence of a 'hole' in the ozone layer over Antarctica, which grew larger at the end of the southern winter. The link between the two phenomena was soon established. Satellite measurements showed that ozone depletion was accelerating, thus reinforcing the first diagnoses of the risks entailed.

In 1987, barely ten years after the alarm was first raised, the Montreal Protocol was signed. Now ratified by 191 countries, this protocol has banned the use of CFC-type gases by industry worldwide, following a timetable adjusted for countries' level of economic development. Under the aegis of the UN, it has set up a monitoring system together with a financial mechanism designed to help industrial reconversion in developing countries. The protocol's successive amendments have regulated the use of substitute gases.

The effect of the Montreal Protocol has been spectacular: CFCs have not been used in industrialized countries since the end of the 1990s (apart from a few very limited medical uses). The last CFCs used in industry by developing countries are banned as from 2010. But has the hole in the ozone layer disappeared as a result? Because of the great chemical stability of CFC gases and the inertia of the atmospheric system, scientists believe that it will require a minimum of fifty years after the last emissions for the hole to close up. The time lag shows how important it is, in situations of this kind, to act early.

If the Montreal Protocol has been such a success, why not adopt its approach in the face of climate risk? In fact this type of regulatory action is not easily transposable to greenhouse gas emissions. CFC gas emissions were confined to just a few thousand industrial installations around the world. Some of these companies, in particular Dupont,

---

[1] The chemists were Paul Crutzen, Mario Molina and Franck Sherwood Rowland.

after constant lobbying to dispute the dangers incurred, developed substitution technologies to replace CFCs, which were ready when the protocol came into force.

Collective action in the face of climate risk began, as in the case of ozone, with the alarm being raised by the scientific community. The rest of the story is rather more convoluted, since much more is involved than regulating the emissions of a few thousand factories worldwide.

The beginning of international cooperation

The first global conference on the climate was held in Geneva in 1979. On this occasion an international scientific research programme was launched, with a view to establishing basic scientific knowledge on the complex and little understood subject of climate change. This programme resulted in the creation in 1988 of the Intergovernmental Panel on Climate Change (IPCC), the scientific panel operating under the aegis of the UN, the importance of which was shown in Chapter 1. The IPCC First Assessment Report came out in 1990. With its publication, a large number of decision-makers discovered that climate risk existed, linked to the accumulation of anthropogenic greenhouse gas emissions. This awareness facilitated the adoption, two years later, of the UN Framework Convention on Climate Change (UNFCCC), which for the sake of simplicity we shall refer to as the Climate Convention.

Signed in 1992 by more than 120 countries at the Earth Summit in Rio de Janeiro, the Climate Convention came into force in March 1994. It has been ratified by 192 parties, nearly all the world's countries. There is only one significant exception: Iraq. The Climate Convention has had less media coverage than the Kyoto Protocol, its main implementation instrument. It is the Climate Convention, however, that lays down the foundations of international cooperation in relation to action against climate change.

The Climate Convention establishes an international governance structure in the face of climate risk. Its leading body is the Conference of the Parties (COP), which brings together the representatives of all states that have ratified the Climate Convention. Its procedures give every country, irrespective of size, an equal voice, in conformity with the principles of the UN. With 192 parties, one can imagine the complexity of the decision-making process and the risk of deadlock or failure to act. Under its statutes, the COP is required to meet once a year, and this usually occurs in the first two weeks of December. The Climate Convention thus establishes a continuous international negotiating process on climate change.

The COP has an operational secretariat based in Bonn, Germany, which is responsible for applying the decisions taken. It also gathers together the information that each party to the Climate Convention is required to provide. This technical work is carried out in association with a subsidiary body responsible for informing the COP on scientific, technological and methodological matters. This body regularly mandates the IPCC, in particular to harmonize greenhouse gas measurement and reporting rules. The secretariat is also responsible for managing the economic mechanisms introduced by the Kyoto Protocol by means of a registry system that counts emissions allowances.

The Climate Convention not only provides a framework for multilateral discussions among countries and an administrative monitoring organization, but it also moves forward the three principles on which international cooperation in the face of climate risk must be based.

The three pillars of the Climate Convention

The Climate Convention's first principle, laid down in its opening statement, is the recognition in international law of the existence of global warming and its link with anthropogenic greenhouse gas emissions. In ratifying the Convention, a state acknowledges the truth of these two propositions. We can nevertheless question the practical import of such acknowledgement. For most of his time in office, President Bush constantly cast doubt on the existence of the link between man-made greenhouse gas emissions and global warming. Yet the US fortunately did not withdraw from the Climate Convention.

The second principle sets out an objective for the international community to cap the concentration of greenhouse gases in the atmosphere. The Climate Convention recommends that the level of this cap be such that it prevents 'dangerous anthropogenic interference' in the climate. If such a level is reached, it should be within a sufficiently long time-frame to allow ecosystems to adapt naturally to climate change; it must ensure that food production is not threatened; and it must allow economic development to proceed in a sustainable manner. Here again, the need to find consensual formulas limits their operational application: neither the time-frame nor the general level at which the cap on greenhouse gas concentrations should be set have been specified.

One can certainly find extenuating circumstances for the Climate Convention negotiators. In 1992, the IPCC had not yet published its First Assessment Report. Decision-makers did not possess as much information as they do today. But as the parties meet each year, they could have used new information that has since been produced by the scientific community to specify the ultimate objective pursued by the Climate Convention. Despite several attempts, particularly in the framework of the 'Shared Vision' working group mandated by the 2007 Bali COP, this has so far not happened.

The third principle of the Climate Convention is that of 'common but differentiated responsibilities' in the face of climate change. 'Common responsibilities' means that in ratifying the Convention, each state agrees to take on part of the collective responsibility. 'Differentiated responsibilities' means that not all states have the same degree of responsibility and that this varies according to their level of development. The differentiation of the degree of responsibility is an equity criterion, the foundations of which can hardly be disputed.

The principle could be applied on a case-by-case basis, in accordance with the parameters characterizing each country's level of development. The Climate Convention gives it a binary interpretation. It classifies countries into two groups: industrialized countries and developing countries. The former, which account for three-quarters of the global greenhouse gas emissions accumulated since 1850, carry a preponderant responsibility. They are listed as Annex I, which includes the member countries of the Organisation for Economic Co-operation and Development (OECD) in 1990 as well as Russia, Ukraine and the countries of eastern Europe. The other 'non-Annex I' countries do not have the same historical responsibility and the Convention recognizes their priority as regards development. The Kyoto Protocol has set in stone this binary interpretation of the differentiation of the degree of responsibility.

The failed ambitions of the Kyoto Protocol

Some COPs have left little trace. Others have marked the history of climate negotiations, as was the case with the COP that met in 1997 in Kyoto. The Kyoto Protocol, adopted in the conference's plenary session, is the main application text of the Climate Convention. It has a twofold impact on international dealings. In legal terms, it introduces a system of commitments into international law that caps greenhouse gas emissions. In economic terms, it matches these commitments with three innovative instruments: the two project-based mechanisms discussed in Chapter 6 and an emissions trading system among countries resulting from the scarcity introduced by the emission caps.

In accordance with the principle of common but differentiated responsibility, Kyoto exempts non-Annex I countries from any quantified commitment to reduce their emissions.[2] To adhere to the protocol, they simply have to commit themselves to conducting an inventory measuring their emissions and periodically providing information on their contributions to collective action on climate change. By signing the protocol, they are able to benefit from the Clean Development Mechanism (CDM), an option that has been widely used by emerging countries, but much less by the least developed countries.

The commitments of Annex I countries cover all anthropogenic greenhouse gas emissions.[3] They apply to all emissions sources located in the national territory of the countries concerned, excluding carbon dioxide (CO_2) emissions arising from international maritime or air transport. These countries are also allowed to offset the storage of carbon by forests and, subject to certain conditions, agricultural land. A complex carbon sink measurement system linked to forests and agriculture has been set up for this purpose.

On average, Annex I countries must reduce their emissions for the 2008–12 period to 5.3 per cent below their 1990 levels. This oddly precise 5.3 per cent figure was a compromise arrived at after long hours of discussion, particularly between the OECD countries and the countries of the former Soviet bloc. Among OECD countries, reductions commitments are 8 per cent for the European Union (EU), 7 per cent for the US and 6 per cent for Japan. Russia, Ukraine and the countries of eastern Europe, which were then in deep recession, were treated generously. They obtained the right to emit annually as

[2] To be exact, we should refer to the Kyoto commitments of Annex B countries, which comprise the Annex I countries of the Climate Convention, minus Belarus and Turkey, which slipped through the net.

[3] As we saw in Chapter 1, these concern carbon dioxide (CO_2), methane, nitrous oxide and three families of fluorinated gases (HFC, PFC and SF6). CFC gases are not covered by the Kyoto Protocol, but come within the Montreal Protocol, the aim of which is to protect the tropospheric ozone layer. Water vapour, the concentration of which varies in the atmosphere independently of human activity, is not covered either.

much greenhouse gas over the 2008–12 period as they were emitting in 1990 – considerably more than their needs, given the fall in their emissions between 1990 and 1997. This surplus of emissions rights, referred to as 'hot air', gave the Russian and Ukrainian governments a substantial security cushion and opened up the prospect of incoming cash flows from the sale of emission allowances. As a result, the Russian parliament ratified the Kyoto Protocol in November 2004, a decision that led to the protocol coming into force three months later, in February 2005.

Matters took a somewhat different turn in the US. The commitment undertaken by Vice President Gore on behalf of the US at Kyoto was very definite. Yet even before his plane took off for Kyoto, the US Senate had unanimously passed a resolution indicating that it would under no circumstances ratify an international treaty imposing a constraint on the US, but leaving emerging countries such as China and India free to emit greenhouse gases.[4] Moreover, President Clinton never sent the text of the protocol to the Senate for ratification.

In March 2001, President Bush clarified the situation by publicly announcing that the US would not be ratifying the Kyoto Protocol. The US exit destabilized the general workings of the Kyoto system. Without the US, Kyoto would cover no more than 40 per cent of global emissions in 1990, the base year, and barely a quarter of emissions in 2005, the year the Protocol began being applied. Furthermore, following the American withdrawal, most Annex I countries, excluding those of the EU, renegotiated a more advantageous calculation of their forest carbon sinks. Economically, the protocol's loss of substance reduced the significance of the international emissions trading market. There was insufficient demand in the face of Russian and Ukrainian 'hot air'. The Kyoto emissions market among nations, which in principle began in January 2008, thus remains largely virtual.

---

[4] The Byrd–Hagel Resolution, named after its two co-sponsors, was passed with a vote of 95–0.

One surprising feature of the Kyoto Protocol is that nothing was said about post-2012 commitments. In accordance with the procedures of the Climate Convention, the choice of rules applying after 2012 was left up to future COPs. Since it requires at least two or three years for an environmental agreement signed within the framework of the UN to come into force, the thirteenth Conference of the Parties, in Bali in December 2007, was made responsible for setting up a discussion process to prepare for the post-Kyoto period. After a somewhat turbulent summit, it managed at the last minute to produce the 'Bali Roadmap', giving a negotiations mandate to two working groups: the first includes the US and is required to develop a shared vision of long-term commitments; the second represents only the 184 countries that ratified the Kyoto Protocol and is tasked with making precise proposals on the commitments for the period succeeding Kyoto. The deadline for these two working groups was the fifteenth Conference of the Parties, held in Copenhagen in December 2009.

Copenhagen: a shift in the centre of gravity of climate negotiations

For two years, several thousand negotiators met within the framework of these two working groups. Some technical sub-groups made progress, especially the one dealing with the question of the protection of forest carbon sinks within the framework of Reducing Emissions from Deforestation and Forest Degradation (REDD). But by and large, the negotiators did not manage to find compromise formulas before the conference. They came up against two main areas of contention: demands by developed countries to extend the commitments to emerging countries; and demands by developing countries for financing in regard to technology transfers and for aid for adaptation to climate change.

This bogging down of the negotiations before the Copenhagen conference was not properly appreciated by the public at large. The situation became worse during the summit, and the UN machinery seemed to jam up further from one day to the next. After a week and a half of negotiations, there was no draft document ready for signing by the heads of state. Their arrival, to great media fanfare, could hardly make up for the accumulated delay. No-one with any experience of climate negotiations was expecting a full agreement on post-Kyoto rules to emerge from the conference under such conditions. The best that could be hoped for was a general policy declaration, needing then to be converted into operational procedures.

This is precisely what happened on the final day of the summit, with the signing of the three-page document known as the Copenhagen Accord. This accord was drawn up during discussions between the representatives of the large emerging countries China, India and Brazil, and the US president. There was a measure of trickery involved: in the UN institutional mechanism, Copenhagen should have allowed a post-Kyoto climate agreement to be developed. Yet none of the countries committed to the Kyoto Protocol contributed to the drafting of the summit's final declaration. After some hesitation, Europe and Japan were finally won over to this text written without them, as too were most developing countries, which were enticed by the promises of additional financing. Despite the opposition mounted by a small group of countries led by Cuba, Kuwait, Ecuador and Nicaragua, the fifteenth COP 'took note' of the Copenhagen Accord, which is a diplomatic way of saying that it was not adopted due to lack of unanimity: in fact only 119 parties out of the 192 making up the COP gave their support to this text.[5]

What do we find in the Copenhagen Accord? To be frank, this very short document could have been signed from the first day of the conference. It restates commitments, most of which had been made public before the conference. These commitments were made by all the developed countries, but also – and this is Copenhagen's

[5] The present work incorporates information available up to the end of April 2010.

main advance – by the large emerging countries. It promises financial transfers to developing countries amounting to $30 billion by 2012 and $100 billion by 2020, without specifying how they will be implemented. The agreement refers to the principles of the 1992 Climate Convention and even gives a figure: the average temperature increase stabilization target should be 2°C. This is the level implicitly recommended by the IPCC, but hitherto adopted only by the EU. Nevertheless, neither the commitments put down in writing in the agreement nor the economic and financial instruments to be set up specify how this very ambitious target is to be achieved.

A failure or a historic advance? The European media, after having blown up out of proportion what was at stake in the Copenhagen conference, often gave a catastrophist view of it. In contrast, the Indian and Chinese media praised the evident success of their respective diplomacy. In actual fact, Copenhagen marks a shift in the centre of gravity of climate negotiation, reflecting the growing weight of the large emerging countries, without which no major decision can be taken. It embodies the big comeback of American diplomacy, which established itself as their favoured interlocutor. It represents a serious setback for Europe, which participated in the summit in a disorganized manner without playing its main trump card in the climate game, i.e., the EU emissions trading scheme (ETS). Finally, the conference underlines the sizing up of the UN institutional framework, and in particular the impossibility of effective climate governance on the basis of the rule of unanimity for the 192 parties to the Climate Convention.

This gravitational shift will allow international cooperation in the face of climate risk to be strengthened subject to three conditions: widening emissions reduction commitments in an effective way throughout the world; refocusing the role of the UN on its tasks of spreading knowledge and monitoring commitments; and setting up economic and financial instruments that are equal to the stakes involved. Let us review each of these elements in turn.

A variable geometry system

The first outcome of Copenhagen was the reinsertion of China and the US into the heart of the commitment system. This is of course crucial: between them they account for a third of world greenhouse gas emissions. The scenario of this return was basically described in the 2003 book *Reconstructing Climate Policy*. Noting the impossibility for the US of joining the Kyoto framework, the authors, Richard Stewart and Jonathan Wiener, advocate a twofold reinvestment of the US in the climate issue: internally, by developing a credible greenhouse gas emissions reduction mechanism; and externally, through diplomacy primarily directed at China and the large emerging countries to enable them to participate in a cap-and-trade scheme intended to combine with the mechanisms stemming from the Kyoto Protocol (Stewart and Wiener 2003).

At Copenhagen, the US committed itself to reducing its greenhouse gas emissions by 17 per cent by 2020, 42 per cent by 2030 and 83 per cent by 2050 compared to 2005 levels. These figures are included in the proposed legislation discussed in Congress, the broad outline of which we presented in Chapter 4. The US commitment is explicitly premised on the adoption of this legislative proposal by Congress, which would enshrine these objectives in American law. The position taken by President Obama at Copenhagen is aimed at avoiding the precedent of Kyoto, i.e., adopting an apparently committing text, but then not being able to get it ratified. China had made public its commitment put in writing in the Copenhagen Accord before the conference, during a visit to Beijing by the US Secretary of State, Hillary Clinton. Unlike that of the US, its commitment is not absolute but relative. It involves reducing by 45 per cent the carbon intensity of its gross domestic product (GDP) (the ratio between emissions and GDP) by 2020 compared to 2005 levels. This commitment is therefore less constraining. If we project 8 per cent annual growth from 2005 to 2020, slightly lower than that of the previous fifteen years, this in fact authorizes an *increase* in emissions of around 75 per cent. The fact remains that for the first time in the history of climate negotiations, a country that is not a member of Annex I of the Climate Convention has made a written commitment in relation to its emissions level.

The general system of commitments arising from the Copenhagen Accord matches those of the US and China. The industrialized countries, grouped together in Annex I of the Convention, have made absolute commitments.[6] The large emerging countries have made relative commitments. We have here a new way of interpreting the principle of 'common but differentiated responsibilities' in the face of climate change.

If we put together the commitments made by Annex I countries, we end up with a figure of around 12 per cent to 18 per cent emissions reduction compared to the base year 1990. Several countries have submitted minimum commitments, with higher targets under a scenario of international collaboration. For example, the EU has committed itself unilaterally to a 20 per cent reduction, while indicating that it is prepared to raise its contribution to 30 per cent in the event of a satisfactory international agreement.

The 12–18 per cent figure falls short of what is generally recognized as being necessary to limit global warming to 2°C. In reality, a number of technical parameters could significantly reduce the scope of the commitments made by the Annex I countries: firstly, the way carbon sinks linked to forests and changes in soil use are calculated; whether or not surplus rights accumulated during the Kyoto period can be used (of relevance here is the Russian and Ukrainian 'hot air' problem); the possibility of double counting if certain countries use offset mechanisms for emissions reductions already counted in

---

[6] This is excluding Turkey and Ukraine, the only two countries in Annex I of the Convention not to have signed the Copenhagen Accord.

another country's commitments.[7] By adding up all these unfavourable effects, one could even end up with a slight increase in the Annex I countries' emissions compared to 1990, instead of the objectives proclaimed. This possibility shows the importance of systems for registering commitments and measuring and verifying outcomes.

The commitments made by large emerging countries concern China, India, Indonesia, Brazil, South Korea, Mexico and South Africa. They are all expressed as relative values, either in GDP carbon intensity as in the case of China and India or with reference to the business-as-usual scenario. How these are made up has not been made explicit in the documents handed over to the secretariat of the Climate Convention. Apart from Brazil, which details its commitments by emission type, what exactly the emissions cover has generally not been specified. Lastly, the fact of setting relative objectives makes a country's right to emit depend on its growth rate: the greater the growth rate, the more it emits, and the converse – a very poor incentive. Moreover, it is not the carbon intensities of GDP or the divergence from the business-as-usual scenario that alters the composition of the atmosphere, but the absolute level of emissions.

The conditions under which emerging countries' commitments can be calculated and verified have been subject to bitter negotiations that have led to an unsatisfactory compromise: the emerging countries have refused to submit to the same reporting requirements to the UN as those applying to Annex I countries. They have accepted such external 'inspection rights' only for measures taken with financial support from industrialized countries. For as long as their commitments are not verifiable by third parties, their credibility will be questionable.

---

[7] For example, in its commitments Russia includes emissions reductions obtained through Joint Implementation (JI) projects of which the offsets, bought by the US or Europe, would also be counted in their emissions reductions.

Confronted by the inability of the COP to further develop the commitment system inherited from the Kyoto Protocol, the Copenhagen Accord introduced a new way of setting targets. This 'variable geometry' system leads to objectives that, because of their many ambiguities, are difficult to assess in terms of what they seek to do. These objectives particularly concern the planet's twelve most highly emitting countries: the so-called G12, which account for three-quarters of human greenhouse gas emissions.[8] This is more than sufficient to pull in the rest of the world if the commitment system constructed is credible and attractive. The G12 comprises both Annex I and non-Annex I countries. It could form the embryo of an authority for managing a future commitment system, on condition that straightforward and transparent rules are established as to the requisite conditions for other countries rapidly to enlarge the circle. A properly functioning climate G12 would be a lot more productive than a G192 bogged down in interminable discussions.

For the G12 to get under way and quickly expand, there needs to be a reliable and impartial mechanism for measuring, registering and verifying commitments. This task should be taken on by UN agencies within the framework of renewed international climate governance.

Post-Copenhagen climate governance

In the immediate aftermath of the Copenhagen conference, international climate governance consists of three superimposed layers: the 1992 Climate Convention and its 192 parties; the Kyoto Protocol and its 184 signatories; and the Copenhagen Accord, supported by 119 countries. Does this growing complexity not risk leading to a totally ungovernable system? That will certainly be the case if a

---

[8] This counts the twenty-seven-member EU as a single country: the reader can find the figures at the end of the book in the table 'Greenhouse gas emissions around the world'.

redistribution of tasks is not implemented and the four pillars that the Climate Convention and the Kyoto Protocol wanted to unite into a single whole are not organized differently. These four pillars are: the commitment system; information on scientific knowledge about climate change; standards and assessment and verification systems; and economic instruments.

We have just seen the ongoing evolution of the commitment system, which has clearly already left the COP and Kyoto frameworks. Let us now examine the next two links, regarding the monitoring capacity of the UN bodies.

This capacity is based on the competences of the IPCC. In addition to its work summarizing and diffusing current scientific knowledge about the climate, the IPCC has the task of establishing the standards that enable greenhouse gas emissions to be calculated in a coherent manner. Although this task might appear to be secondary, it is in fact of prime importance. The whole edifice of the carbon economy is founded on the reliability and homogeneity of measurements. And it can be clearly seen what happens if there is no impartial authority to lay down reliable and generally known measurement rules: each party permanently haggles over its rights, twisting the calculation methods to its own ends. Such, by and large, is what has happened in regard to counting forest carbon sinks in the national inventories of Annex I countries since 2001. As we have also just seen, the interpretation of the Copenhagen Accord remains subject to persistent vagueness in this respect.

It is thanks to the system of standards constructed and validated by the IPCC that national greenhouse gas inventories can be drawn up. It is a matter here of compiling statistical tables that enable each country's emissions to be calculated. Without such inventories, it is strictly impossible *ex ante* to put precise figures on each country's objectives, let alone verify *ex post* how far these have been attained. For this reason the setting up of national inventories, carried out every year and sent to the secretariat of the Climate Convention, is a rule that should be imposed on all countries wishing to join international climate agreements. For the majority of countries outside Annex 1, this is at present far from being the case. The consolidation of information, in association with the national environmental agencies that produce it, would allow emissions throughout the world to be monitored more effectively.

The second pillar of the mechanism for monitoring climate agreements is a system of registries in which emissions allowances are recorded. Chapter 4 showed how important these registries are for the functioning of the European emissions trading market. At a global level, the UN registry system enables the commitments made by various countries to be recorded, avoiding in particular any double counting, and their implementation to be tracked. The UN registry is also an indispensable instrument for the UN specialist agencies responsible for validating emissions reductions associated with projects and the issuing of the corresponding offsets. The existence of a reliable, secure registry system is a necessary condition for setting up the economic instruments, without which no climate agreement has the slightest chance of being implemented in practice.

The linchpin of the economic instruments

In an international climate agreement, economic instruments have a twofold role: encouraging economic players to reduce their emissions at the lowest cost, through the carbon price; and finding a way of distributing the carbon rent created by this price so that it gives countries an incentive to join the agreement.

The Kyoto Protocol was constructed with the intention of putting a carbon price on countries' rights to emit, but without really anticipating the distributive effects that could result from the introduction of such a price. As the previous chapter showed, the grandfathering rule used for the initial allocation and the absence of a long-term target did not give developing countries any incentive to join the agreement.

In fact it could even encourage them to increase their emissions so as to gain more rights when they possibly joined later, or through the CDM if this was not properly managed. This first construction flaw led to the stalling of the negotiations on widening commitments in the framework of Kyoto. The mechanism's second failing is to have tried to create a carbon market among sovereign states without first having a public authority over them to set the rules and impose arbitration decisions. Such a system, though attractive in principle, can only function in practice once the constraints become effective: if an industrial installation is not in compliance in the EU ETS, it costs it €100 per allowance not surrendered, to be paid immediately and without redress. In the Kyoto system, as soon as a constraint arises for a country, it launches into negotiations around the calculation of its rights as regards forest carbon sinks.

Will the Copenhagen Accord succeed in overcoming this obstacle inherited from the Kyoto architecture? One of its characteristics is the lack of any precise reference to economic instruments. This absence leaves considerable doubt as to the future of the Kyoto Protocol's flexibility mechanisms. This is not so much a problem for the system of emissions trading among countries, which is and, in all likelihood, will remain virtual. But it is more worrying in regard to the two project-based mechanisms, which sustain the flow of low-carbon investments; and this flow risks drying up in the absence of visibility on the valorization of offsets after 2012.

The substance of the Copenhagen Accord will in fact depend on the capacity of the signatory countries to set up economic instruments and to ensure that these are quickly and optimally coordinated. The advantage of this bottom-up method is that such instruments already exist: they are the cap-and-trade schemes described in Chapter 4 and the project-based mechanisms presented in Chapter 6. The problem is that this process is dependent on the rate at which they can be deployed in each country. Let us look briefly at the conditions within the G12 countries, consisting of the world's twelve largest emitters.

To facilitate the attainment of the Copenhagen commitments up to 2020, and especially to prepare for the acceleration in emissions reduction in subsequent periods, it is essential first of all to work on energy-related emissions. The prototype EU ETS covers slightly more than 4 per cent of global CO_2 energy-related emissions. The second act will be played out in the US Congress and will determine what comes next: until such time as Congress introduces legislation controlling emissions in the US, the Copenhagen Accord will remain at a standstill. Once the necessary laws are voted through, the proportion of energy-related emissions covered by cap-and-trade is likely to increase by a factor of 2.5 to 3.5, depending on the details of the legislation adopted. Real climate negotiations can then begin.

An American change of heart will unquestionably make life easier for the Australian government, which had already twice attempted to pass legislation capping the country's emissions, and will speed up the coverage of Japanese energy-related emissions which are already partially capped. Canada, which has made the same Copenhagen commitments as the US, would almost certainly come on board too. At this point we have a much more substantial foundation in terms of emissions covered, especially if Europe manages to widen its carbon pricing by introducing taxation of CO_2 in transport and diffuse emissions. With some 11 to 15 billion tonnes of CO_2, the greater part of the industrialized countries' energy-related emissions would then be capped (or priced by means of a tax for some diffuse emissions). The time would then be ripe to enter into serious discussions with the large emitters outside Annex 1.

Of these, the top energy-related CO_2 emitters are China, India, South Korea and Mexico. Over the last fifteen years these four have doubled their energy-related CO_2 emissions. They present very different levels of development. Korea, a member of the OECD and now wealthier than many European countries, remains in this group due to a classification error derived from the past. At the other extreme, India is a country which is only just (but very quickly) leaving the group of least developed countries in the world. These developing countries have to be made partners to energy-related carbon pricing in accordance with their own specificities. The cap-and-trade mechanism for energy-related emissions is an excellent negotiating instrument that can have considerable appeal for these countries and lead them to clarify and deepen their commitments in regard to this emission category. Negotiations could focus on what is covered by capped emissions, the initial level of caps and their evolution over time, the degree of openness to external allowances markets, the type of initial allocation and how the carbon rent will be used once it is created. Needless to say, there will be some hard bargaining. But the rulers of emerging countries have a threefold interest in their successful outcome. Firstly, cap-and-trade is the least costly way for them to participate in the collective effort to reduce greenhouse gas emissions. Secondly, it opens up part of their economies to a new market that encourages them to get ready for the low-carbon standards that will make tomorrow's industries competitive. And thirdly, integrating an energy-related emissions trading mechanism is an operation that can quickly procure additional revenue for their budgets. This can be achieved through auctioning a proportion of the allowances from capping industrial emissions and through financial compensations that they will undoubtedly negotiate to join the system.

Brazil and Indonesia are intermediate-size emitters of energy-related CO_2, but if their agroforestry emissions are taken into account then they are the fourth and fifth largest greenhouse gas emitters in the world. It is with these two countries that future agroforestry mechanisms need to be tried out on a real-life scale. A persistent idea circulates among climate economists that reductions can be obtained at a low cost by slowing the rate of deforestation. However, the situation is more complex than that of energy-related emissions, for there is no prototype that has already been subject to large-scale testing. Moreover, only Brazil possesses a system for emissions measurement and verification, and without these little more can be constructed than what has been done already in the framework of voluntary offsets.

Negotiating with Brazil is also important because it has become one of the largest producers and exporters of agricultural products on the world market. It is therefore impossible to bring it into forest carbon pricing without directly including the agricultural dimensions and the forest issue.

There remains one member of G12 that we have not yet discussed: the Russian Federation. Russia is a member of Annex I and has ratified the Kyoto Protocol. It is officially fully integrated into the UN climate commitment system. In the Copenhagen Accord it committed itself to reduce its emissions by 15–25 per cent compared to 1990 levels, or a potential increase of 17–33 per cent compared to 2005 levels. In view of the flexibility it made a point of negotiating as regards forests and the 'hot air' stock inherited from the past. In fact Russia is not subject to any constraint. More seriously, like other large producers and exporters of hydrocarbons, it has no incentive to join a constraining agreement, the success of which would in the long term eat into its oil and gas rent. It is therefore essential to start discussions with this group of countries on how to include them in the joint management of carbon and oil rents, along the lines sketched out in Chapter 7 in relation to the carbon sink allocation model.

In terms of economic instruments, the Copenhagen Accord reverses the situation by changing from Kyoto's top-down architecture to a bottom-up system. This new approach does not reduce the short-term uncertainties but opens up real longer-term prospects if the major emitters succeed in negotiating good arrangements. The successful deployment of economic instruments is essential for achieving emissions reduction commitments. In addition, it lends credibility to the pledges of financial transfers made in Copenhagen.

Mobilization of financing

At the financial level, the Copenhagen Accord endorses the principle of a financial transfer of $30 billion a year by 2012 and of $100 billion a year by 2020 to strengthen the adaptation capacities of countries that are most vulnerable to climate change and to stimulate low-carbon technology transfers to developing countries. But how is $100 billion a year to be transferred from the global North to the global South? It is not enough to plan on setting up a fund and calling it the Copenhagen Green Climate Fund. It is necessary to define the rules for sharing it among beneficiaries and the processes ensuring that the funds are properly used. Above all, $30 billion, then $100 billion a year has somehow to be found.

Let us say straightaway, there is practically no chance in the near future that such sums will come from the budgets of the industrialized countries. One of the consequences of the crisis was lasting impoverishment of public finances. To find financing over and above what already exists, three sources must be tapped.

The most direct is the growth of carbon offsets, which allows, through the purchasing of credits, emissions reduction investment to be financed in developing countries. The two most immediate conditions for ensuring this growth are, on the supply side, the development of programmatic methodologies and, on the demand side, the extension of the cap in order to create a sufficient outlet for carbon offsets. An important issue in terms of governance is the reinforcement of project standardization under the authority of the UN. The alternative would be to let the various carbon markets compete, with each of them seeking to attract the best projects. But apart from being needlessly expensive, this alternative would in particular give little chance to small projects.

By increasing carbon rent, the widening of energy-related emissions constraints opens up a second potential source of transfers to developing countries: a proportion of the proceeds from the sale of allowances by governments to industry could be recycled towards developing countries.

There is, finally, a third source of financing that has considerable potential: the long-term financial assets of institutional investors, including pension funds, sovereign wealth funds and large insurance companies. What such long-term investors all have in common is managing well-endowed financial portfolios with the aim of increasing their value over several decades – the requisite time scale for the transfers aimed at by the Copenhagen Accord. Some of these actors, such as the Caisse des Dépôts in France or KFW in Germany participated in the launch of the carbon market, allocating funds to purchasing offsets under the CDM or Joint Implementation (JI). In the same vein, it is a Norwegian fund that currently provides the greater part of transfers in relation to experiments in avoided deforestation. A clear signal by large emitters of their wish to see a long-term carbon price emerge would help increase this type of investment that until now has remained relatively limited considering the sums mobilized. Some sovereign wealth funds are used, moreover, for recycling oil rent – an additional reason for mobilizing them, since this is an additional way of enabling carbon rent and oil rent to coexist more harmoniously in future.

But it is not enough simply to raise money. It has to be converted into effective projects that achieve their targets: the adaptation of countries most vulnerable to climate change and the transfer of low-carbon technology to those who really need it. The carbon price is an instrument that can help attain the second of these. It is of no use for converting financing into projects that strengthen the capacity of the 300 million people living in Asian deltas or the hundreds of millions of farmers in subtropical zones to confront the tightening of constraints that climate change imposes on them.

Faced with the issue of adaptation to climate change, the Copenhagen Accord puts forward an excellent principle: solidarity among countries to help the most vulnerable. But putting this principle into practice is complicated, for we do not currently have the instruments to give expression operationally to such new solidarity. The forms of governance for achieving it cannot be limited to the inner circle of the planet's large greenhouse gas emitters, on which the arrangements liable drastically to alter the world's emissions trajectories depend. Most other countries must be involved as well. This is the most difficult issue to resolve for the credibility of any climate pact that Copenhagen may lead to.

What is the credibility of a climate agreement based on?

Like the action successfully taken through the Montreal Protocol against the destruction of the ozone layer, action in the face of climate risk was triggered by the scientific community raising the alarm. The credibility of the Montreal Protocol was based on a legally constraining instrument applied to a limited number of emissions sources. Substitution technologies were available and a relatively simple financial instrument enabled these to be transmitted to developing countries. The credibility of a collaborative approach to reduce greenhouse gas emissions requires an alternative basis, because of the large number of emissions sources and because of economic and technological stakes of an entirely different order.

The Kyoto Protocol, the main implementation instrument of the Climate Convention, which establishes general principles of action, was based on a legally binding constraint on the emissions of developed countries. On the basis of this constraint, economic instruments were developed, the linchpin of which should have been an ETS between states, generating a carbon price. This system was not able to achieve its objectives, partly due to the withdrawal of the US, but also because of shortcomings in its construction that limited the effective deployment of economic instruments.

The Copenhagen conference led to a profound change of great magnitude. Initially planned to devise rules as a follow-up to the

Kyoto Protocol, it resulted in a political agreement drawn up by countries not committed to this protocol. This overall shift puts the emerging countries and the China–US couple – which accounts for a third of all anthropogenic greenhouse gas emissions – at the centre of negotiations.

The credibility of this political agreement does not rest on legally binding commitments set down in the framework of an international treaty. Rather it depends on the capacity of the large emitters to coordinate the deployment of economic instruments regulating and pricing their domestic energy-related emissions. Such pricing is based on known instruments that have been tried and tested in Europe and in the context of the Kyoto Protocol's project-based mechanisms. It will be easier to deploy if the UN bodies concerned reinforce their registration and verification system for projects.

Copenhagen also introduces the principle of financial offsetting intended for those countries most vulnerable to climate risk. In a situation where budgets are likely to remain tight for the foreseeable future, sticking to this pledge will involve recycling a proportion of carbon rent, which in turn calls for extensive coordination of national climate policies.

Setting up these economic and financial systems will require handing over the management of future climate agreements to a multilateral body with sound economic and financial skills, in which the interests of all countries, not only those of the large emitters, will be properly represented. In terms of governance, this entails entering uncharted waters.

Both action and inaction entail risks

The light is fast fading. The north wind is bitingly cold. But the chill does not discourage the many cyclists pedalling along the city streets. At the opera house by the town's docks they're getting ready for the evening performance. Behind its imposing roof, the last glimmer of daylight is reflected from the blades of the wind turbines offshore. We are in Copenhagen, a city where the environment is taken seriously.

A ship is inching its way under Oresund Bridge, which links Copenhagen to the Swedish city of Malmö. It is a bulk carrier, transporting coal loaded in Szczecin, Poland. It glides along the stretch of the Baltic Sea bordering the city, avoiding the wind turbines lined up like marker buoys in the water and berths behind the opera house. Its cargo is destined for Amagervaerket power station, situated midway between the opera and the wind farm.

The whole city is brightly lit. Light shining through the huge glass wall of the opera house casts multi-coloured glints across the water. Plumes of water vapour emerge from the three chimneys of the power station, now running at full capacity to meet peak demand for electricity.

From the Bella Center, the conference centre where the Copenhagen summit was held in December 2009, one can clearly make out the juxtaposed wind farm and thermal power station, symbols of tomorrow's and yesterday's worlds. What the negotiators discussed above all else were the options countries have in making the transition from yesterday's world to tomorrow's.

Choosing inaction in the face of climate change is to have confidence in the spontaneous reactions of the economic system inherited from the past and in traditional political judgement. This business-as-usual option is generally justified by the uncertainty and supposed cost of action to confront climate change. It deserves to be debated calmly. It also entails risks. Such risks are difficult to quantify because they depend on chain reactions of systems about which our knowledge is imperfect: firstly, the climate system, the retroactive effects of which can amplify or cushion the shock caused by the accumulation of human greenhouse gas emissions; secondly, ecosystems, where the reactions to climate change impacts are observed more easily than they are predicted; and thirdly, economic and social systems, whose resilience to climate shock extending over several decades and probably into the next century is inherently unknowable. Economists have attempted to convert the risks arising from inaction into economic costs. In view of the above-mentioned uncertainties, the figures they come up with are hard to use. Lastly, the breakdowns in economic life that climate shock is liable to provoke have potentially very dangerous consequences. As Harald Welzer (2009) warns us in his book, *Les guerres du climat, pourquoi on tue au XXI° siècle*, climate-related crises could degenerate into new armed conflicts, with extreme violence becoming the way of managing them.

The choice of action involves integrating climate risks, that will weigh mostly on future generations, into our current decision-making. There are, of course, many ways of doing so, with very different implications on the functioning of societies. The approach that we recommend in this book draws upon economic mechanisms that have been tested in real-life situations thanks to the first steps taken by international climate diplomacy: the European emissions trading scheme and the project-based mechanisms inherited from the Kyoto Protocol. The difficulty with this approach is coming up with compromises that can appeal to the main groups of countries, with their often divergent short-term economic and energy interests.

Let us now weigh up the respective risks of action and inaction. The risk associated with action is that of wasting resources by overestimating the climate threat. Suppose that through favourable accumulated retroactive effects, future climate change impacts are much more limited than what most scenarios offered by climatologists lead us to believe. We would then be in the situation that climate sceptics insist will occur however societies behave. For all that, would the resources committed in order to act in a preventative way on climate change be an enormous economic waste?

If well managed, choosing to act would not be cause for regret, since it would result in embarking upon two radical changes that our societies in any case greatly need. What risk do we take in too rapidly reducing the dependence of the energy system on oil, gas and coal? Such dependence weakens the economy as a whole, as was shown by the deep recession of 2008–9. It has, moreover, become one of the main sources of armed conflict in the world. What risk do we take by committing too many resources to step up agricultural development in order to reduce food insecurity while protecting the capital constituted by tropical forests? Here too, excessive action against climate change cannot lead to any regrets. In both cases, taking account of climate change leads in fact to tackling major questions to which the current functioning of society offers no satisfactory answers. It is for this reason that the climate can become one of the key elements in the historical reorientation that the capitalist system must inevitably undergo as a result of the economic and financial crisis.

Globalized financial capitalism opened up national economies and has been the source of an unprecedented wave of progress in the emerging world. But this wave is coming up against limits that the markets, as currently organized, are incapable of revealing, still less of managing. What gave rise to the crisis? Must we now abolish markets or drastically reduce their influence by extending regulation? Let us listen to what one of the great specialists in economic development, the Nobel prize winner Amartya Sen (1999: 191) says: 'The problems, and they exist, usually have other causes than the existence of the market as such [...] One does not settle these problems by getting rid of markets, but by allowing them to function better and more fairly.'

The price of carbon is a signal resulting from a political decision that encourages resources to be shifted through modifying the system of benchmark prices. By taking account of the scarcity of the atmosphere in the scale of economic values, it greatly improves the functioning of the markets. It prefigures a system in which the costs and risks of assaults on the environment will be correctly priced. Without taking this into account, the opening up of the markets in recent decades will not be able to be maintained and the contraction of the economic sphere rapidly risks undermining the advances in living standards that it generated.

The payoff for pricing carbon is the creation of carbon rent, the distribution of which is central to the question of equity raised by Sen. The disproportionate weight of the large historical emitters resulted in the adoption of the grandfathering model for setting emission caps. This model is advantageous to them, but is unsatisfactory from the standpoint of equity. To bring carbon pricing into general use in the economic system, it is advisable to combine this model reflecting yesterday's world with the model aimed at an egalitarian sharing out of rights. It is this issue of a system of long-term objectives that climate negotiators have found hard to reach agreement on. But in the short term this makes little difference to the poorest countries, which are least responsible for and most vulnerable to climate change. Hence equity, as much as efficiency, calls for the transfer of part of the carbon rent to these countries so as to immediately strengthen their capacity to adapt. The Copenhagen Accord recognizes the principle of such a transfer, but in terms that are too general to be useful. Future progress in climate negotiations depends to a great extent on the negotiators' ability to give it operational force.

The inclusion of equity lies at the heart of the climate risk issue. It has two aspects: equity within the present generation; and equity between the present generation and those that will succeed and inherit a world shaped by our immediate choices. The question of equity concerns economic organizations, but also comes back to our own values, which stem from our way of representing the world. At this point economists come up against the limits of their discipline and have to turn to the teachings of philosophers.

A major debate in the history of Western thought arose in the seventeenth century, with the opposed world views of Descartes on the one hand and Spinoza on the other. In many respects we are the heirs to Descartes, in that he taught methodical doubt, scientific method and self-awareness. Descartes' thinking focussed on the individual subject, who is able to dispel doubt through the introspective approach – the famous *Cogito ergo sum* – and arrive at knowledge of the external world, which he is able to make better use of thanks to his knowledge. This ethnocentric view of nature, dominated by mankind, was then systematized by the encyclopaedists, during the Enlightenment.

In the Spinozan view, mankind is no longer at the centre, but is rather one link in what Spinoza calls Substance. Progress no longer consists in taming the external world through knowledge but seeking harmony between mankind and the other components of Substance. This view prefigures the question of sustainability confronting our societies right now. Faced with climate risk, should we not simply let ourselves be guided by the precepts of the author of *Ethics*? Before bringing the present work to a close, let us stop for a moment at one of his assessments of what our system of values should be: 'Under the guidance of Reason we desire a greater future good before a lesser present one, and a lesser evil in the present before a greater in the future' (Spinoza, n.d.).

References

Agrawala, S. and Fankhauser, S. (2008) *Economic Aspects of Adaptation to Climate Change: Costs, Benefits and Policy.* Paris: OECD. Available at www.oecd.org/env/cc/adaptationeco.

Aoun, M-C. (2009) 'Middle East and North Africa: Oil Curse or Blessing?' In Jean-Marie Chevalier (ed.) *The New Energy Crisis.* Basingstoke: Palgrave Macmillan.

Chevalier, J-M. (2009) *The New Energy Crisis.* Basingstoke: Palgrave Macmillan. (Originally published in French under the title *Les nouveaux défis de l'énergie.* Paris: Economica.)

Chichilnisky, G. and Heal, G. (1998) *Environmental Markets, Equity and Efficiency.* New York: Columbia University Press.

Chomitz, K. M. (2006) *At Loggerheads? Agricultural Expansion, Poverty Reduction and Environment in the Tropical Forest.* Washington, DC: World Bank.

Ellerman, A. D., Convery, F. and de Perthuis, C. (2010) *Pricing Carbon.* Cambridge: Cambridge University Press.

Food and Agriculture Organization (FAO) (2009) *The State of Food Insecurity in the World 2009.* Rome: FAO.

Godard, O. (2008) *Unilateral European Post-Kyoto Climate Policy and Economic Adjustment at EU Borders.* École Polytechnique, Paris. Sustainable Development Chair. Cahier no. DDX 07-15.

Hardin, G. (1968) 'The Tragedy of the Commons', *Science,* **162:** 1,243–8.

Houghton, J. T. (2010) *Global Warming: The Complete Briefing.* Fourth edition. Cambridge: Cambridge University Press.

Intergovernmental Panel on Climate Change (IPCC) (2007) *Climate Change 2007, 4th Assessment Report.* Cambridge: Cambridge University Press.

Maddison, A. (2001) *The World Economy: A Millennial Perspective*. Paris: OECD Development Centre.

Sen, A. (1999) *Development as Freedom*. Oxford: Oxford University Press.

Shapouri, S., Rosen, S., Meade, B., and Gale, F. (2009) *Food Security Assessment 2008–9*. Washington: US Department of Agriculture.

Spinoza, B. (n.d.) Ethics, Part IV, Proposition LXVI.

Stewart, R. B. and Wiener, J. B. (2003) *Reconstructing Climate Policy: Beyond Kyoto*. Washington, DC: The American Enterprise Institute Press.

Stoft, S. E. (2008) *Carbonomics: How to Fix Climate and Charge it to OPEC*. Nantucket, MA: Diamond Press. Available at: http://stoft.com.

Tirole, J. (2009) *Politique climatique: une nouvelle architecture internationale*. Report to the Conseil d'Analyse Economique, La Documentation française.

Tol, R. S. J. (2005) 'Adaptation and Mitigation: Trade-offs in Substance and Methods', *Environmental Science and Policy*, 8: 572–8.

Wara, M. W. and Victor, D. G. (2008) *A Realistic Policy on International Carbon Offsets*. Stanford University, working paper no. 74, April.

Weitzman, M. L. (1974) 'Prices versus quantities', *Review of Economic Studies*, 41: 447–91.

Welzer, H. (2009) *Les guerres du climat, pourquoi on tue au XXI° siècle*. Paris: Gallimard.

APPENDIX I

Thirty key readings

1. Agrawala, S. and Fankhauser, S. (2008) *Economic Aspects of Adaptation to Climate Change: Costs, Benefits and Policy*. Paris: OECD. Available at: www.oecd.org/env/cc/adaptationeco.

    The authors emphasize the fragmentary nature of knowledge about the cost of adaptation policies to climate change. They warn about the unreliability of regularly used figures and suggest ways of improving economic instruments.

2. Aldy, J. E. and Stavins, R. N. (2009) *Post-Kyoto International Climate Policy: Implementing Architectures for Agreement*. Cambridge: Cambridge University Press.

    Arising initially from a seminar held at Harvard in the spring of 2006, the second edition of this collective work has been considerably expanded. The editor sensibly provides a short version of what has become a standard reference work in American universities.

3. Bellassen, V. and Leguet, Benoît (2008) *Comprendre la compensation carbone*. Paris: Pearson.

    Well-known as specialists in the field of carbon offsets, Valentin Bellassen and Benoît Leguet provide a very helpful guide to the economics of emissions reduction. For the first time, the rules of the Kyoto Protocol are explained in simple and straightforward language.

4. Broecker, Wallace S. and Kunzig, Robert (2008) *Fixing Climate: What Past Climate Changes Reveal about the Current Threat – and How to Counter it*. New York: Hill and Wang.

    Wallace Broecker was one of the first scientists to reconstruct the history of past climates. With the journalist Robert Kunzig, he provides a new view on climate risks and how to reduce them. An acerbic interpretation.

5. Chevalier, J-M. (2009) *The New Energy Crisis*. Basingstoke: Palgrave Macmillan. (Originally published in French under the title *Les nouveaux défis de l'énergie*, Economica, 2009).

In this fascinating book, the team of the *Centre de Géopolitique de l'Energie et des Matières Premières* at the University Paris-Dauphine under the coordination of Jean-Marie Chevalier addresses all the main energy questions. Links to the climate are made throughout.

6. Chichilnisky, G. and Heal, G. (1998) *Environmental Markets, Equity and Efficiency*. New York: Columbia University Press.

This relatively technical work enriches the analysis of the functioning of emissions allowances markets by incorporating the equity conditions resulting from allocation choices.

7. Chomitz, K. M. (2006) *At Loggerheads? Agricultural Expansion, Poverty Reduction and Environment in the Tropical Forest*. Washington, DC: World Bank.

A key reference for anyone wishing to understand the complexity of the mechanisms related to agriculture and tropical deforestation. This report, accompanied by high quality illustrations, is written straightforwardly and clearly.

8. Delbosc, A. and de Perthuis, C. (2009) *Idées reçues sur le changement climatique*. Paris: Le Cavalier Bleu.

This book aims to help readers situate themselves in regard to filtering received ideas on climate change. It owes much to the pedagogical talents of Anaïs Delbosc and her insistence on constructive formulation.

9. Dessler, A. E. and Parsons, E. A. (2006) *The Science and Politics of Global Climate Change: A Guide to the Debate*. Cambridge: Cambridge University Press.

The authors are specialists in the ozone layer. They provide insightful clarification on the place of science in political debate. This compact book is one of the best introductions to the subject.

10. Ellerman, A. D., Convery, F. and de Perthuis, C. (2010) *Pricing Carbon*. Cambridge: Cambridge University Press. (French edition: *Le prix du carbone*, Paris: Pearson.)

This book closely analyzes the European emissions trading market. It resulted from a three-year collaboration between European and American researchers centred on the 'Parisian circle' then based at the Mission Climat of the Caisse des Dépôts. My intellectual debt to my two co-authors is immense.

11. Godard, O. (2008) *Unilateral European Post-Kyoto Climate Policy and Economic Adjustment at EU Borders*. École Polytechnique, Paris. Sustainable Development Chair. Cahier no. DDX 07-15.

Olivier Godard is one of the leading French economists concerned with climate issues. He here examines the conditions that a border adjustment mechanism must meet to avoid carbon leakage.

12. Hardin, G. (1968) 'The Tragedy of the Commons', *Science*, 162: 1,243–8.

Like a fine Bordeaux wine, this paper improves with time. Reading it was for me a revelation and a source of inspiration for writing Chapter 8.

13. Helm, D. and Hepburn, C. (2009) *Economics and Politics of Climate Change.* Second edition. Oxford: Oxford University Press.

In this second collective work on the climate, Dieter Helm is joined by another researcher at Oxford University, Cameron Hepburn. The book gathers together some of the leading names in the economics of climate change. A very comprehensive compendium.

14. Houghton, J. T. (2010) *Global Warming: The Complete Briefing.* Fourth edition. Cambridge: Cambridge University Press.

Professor of atmospheric physics at Oxford University, former chief executive of the UK Meteorological Office and co-chair of the first IPCC scientific assessment working group, the author has all the qualities to write the definitive work on climate change. He has done so in simple and easily accessible language. Essential reading.

15. International Energy Agency (IEA), *World Energy Outlook*, published in November each year.

The IEA brings high-level expertise to climate change, and this publication has become the main reference for differentiating energy scenarios up to 2030. Another useful IEA publication is *Energy Technology Perspectives*, covering the period up to 2050.

16. Intergovernmental Panel on Climate Change (IPCC) (2007) *Climate Change 2007, 4th Assessment Report.* Cambridge: Cambridge University Press. The summary report and three workshop reports are published by Cambridge University Press. Available at: http://www.ipcc.ch/.

An essential reference work on the state of climate science. But not short: more than 3,000 pages in all. Begin with the summary report, which runs to less than 100 pages.

17. Jouzel, J. and Debroise, A. (2004) *Le Climat: jeu dangereux.* 'Quai des sciences' collection. Paris: Dunod.

A world-renowned glaciologist, Jean Jouzel is also vice-president of the IPCC no. 1 working group. He linked up with Agnès Debroise, a science journalist, to co-write this clear, accessible synthesis with great policy-making relevance.

18. Le Treut, H. (2009) *Nouveau climat sur la terre, Comprendre, prédire, réagir, Flammarion.* Paris: Nouvelle Bibliothèque Scientifique.

Climate science expounded for non-specialists by one of the leading French climatologists. Staying well clear of media hyperbole, the book offers an objective discussion on the state of knowledge of global warming and the uncertainties that remain. Deserves to be widely read.

19. Mendelsohn, R. (2006) 'The Role of Markets and Governments in Helping Society Adapt to a Changing Climate', *Climatic Change*, 78: 203–15.

This paper proposes a division of labour as regards adaptation: the market concerns itself with producers; governments with biodiversity and public health; and public/private partnership with water management.

20. Nordhaus, W. (2008) *A Question of Balance: Weighing the Options on Global Warming Policies.* New Haven, CT: Yale University Press.

    William Nordhaus is one of the pioneers in the economics of climate change. His work extends back more than thirty years. This experience confers a certain authority for criticizing the Stern Review, a study that has met with far from unanimous acclaim in American academic circles.

21. Stern, N. (2007) *The Economics of Climate Change. The Stern Review.* Cambridge: Cambridge University Press. Available at: www.hm-treasury.gov.uk.

    This work has a similar relation to climate economics as the IPCC report to climate science: the most comprehensive synthesis among existing studies. But be warned: it runs to more than 700 pages.

22. Stewart, R. B. and Wiener, J. B. (2003) *Reconstructing Climate Policy: Beyond Kyoto.* Washington, DC: The American Enterprise Institute Press.

    Written by two eminent jurists, this book considers the options for US climate policy after its withdrawal from the Kyoto Protocol. The importance of the US/China axis is given an outstanding treatment.

23. Stoft, S. E. (2008) *Carbonomics: How to Fix Climate and Charge it to OPEC.* Nantucket, MA: Diamond Press. Available at: http://stoft.com.

    Steven Stoft is a recognized specialist in electricity markets. In this pertinent book, he recommends introducing an 'untax' on carbon in the US. By this he means a tax from which the revenue is fully refunded to households in an egalitarian way.

24. Tirole, J. (2009) *Politique climatique: une nouvelle architecture internationale.* Report to the Conseil d'Analyse Economique, La Documentation française.

    In this report, the author skilfully revisits the conditions for an optimal climate agreement and criticizes the failings of existing systems. Of particular note is the interesting contribution by Christian Gollier on uncertainty.

25. Tol, R. S. J. (2005) 'Adaptation and Mitigation: Trade-offs in Substance and Methods', *Environmental Science and Policy*, 8: 572–8.

    A professor at the University of Hamburg, Richard Tol advocates decentralized adaptation strategies. In this provocative paper, he puts forward the idea that poor countries may suffer as a result of emissions reduction actions taken by the countries of the global North.

26. Trotignon, R. (2009) *Comprendre le réchauffement climatique.* Paris: Pearson.

    A researcher at the Programme de Recherche sur l'Economie du Climat (PREC) at University Paris-Dauphine, the author has the knack of bringing figures to life. In this essay, he provides the reader with key orders of magnitude and comes up with an entirely objective diagnosis. An introduction to the subject, accessible to everyone.

27. Wara, M. W. and Victor, D. G. (2008) *A Realistic Policy on International Carbon Offsets*. Stanford University, working paper no. 74, April.

An uncompromising critique of the functioning of the Clean Development Mechanism (CDM) by two brilliant, if sometimes rather caustic, authors. The article has had a certain influence in heightening suspicion of project-based mechanisms.

28. Weitzman, M. L. (1974) 'Prices versus quantities', *Review of Economic Studies*, 41: 447–491.

Probably the most technical reference in our selection. But this paper is indispensable for those who wish to understand the economic debate between partisans of taxation and partisans of negotiable allowances markets. For initiated readers.

29. Welzer, H. (2009) *Les guerres du climat, pourquoi on tue au XXI° siècle*. Paris: Gallimard.

From the Holocaust to Rwanda, the author is a specialist in extreme violence. His thesis is that climate change risks perpetuating this type of violence. His argument is, regrettably, convincing. Translated into French from German.

30. World Bank, *State and Trends of the Carbon Market*, World Bank, annual publication. Available at: www.wds.worldbank.org.

This annual publication makes reference to carbon finance. Its launch in 2004 and its subsequent success owes much to two French economists, Franck Lecocq and Philippe Ambrosi.

In addition, the 2010 edition of the annual report on development around the world, *Development and Climate Change*, is a mine of information.

Thirty key sets of figures

By Raphaël Trotignon[1]

The atmosphere and greenhouse gases

1. The atmosphere is a thin layer of gas that becomes denser nearer to the ground. When we travel by air (at an altitude of 10 kilometres), around 70 per cent of the molecules comprising it, including greenhouse gases, are below us. At an altitude of 50 kilometres, one has passed through the ozone layer. Beyond that, the atmosphere gives way to empty space.

2. The Earth's atmosphere consists largely of nitrogen (78 per cent) and oxygen (21 per cent). Greenhouse gases (excluding water vapour) account for less than 0.04 per cent of the total. These percentages of the total volume exclude water vapour, the concentration of which (0–4 per cent) depends to a great extent on weather conditions.

3. The various greenhouse gases do not remain indefinitely in the atmosphere. But how long each gas stays there varies. A molecule of methane remains in the atmosphere for twelve years on average, a molecule of nitrous oxide 114 years, and industrial gases from hundreds up to thousands of years. Due to its very complex cycle, carbon dioxide (CO_2) can remain there between two years and several thousand years.

4. Not all greenhouse gases have the same warming power. Over 100 years, a tonne of methane (CH_4) will heat the atmosphere twenty-five times as much as

[1] Raphaël Trotignon is a researcher at the Climate Economics Chair (CEC) at University Paris-Dauphine. He is the author of *Comprendre le réchauffement climatique* (2009).

a tonne of CO_2. The same quantity of nitrous oxide (N_2O) will heat it 300 times as much, and industrial fluorinated gases (HFCs, PFCs, SF_6) several thousand times as much. The tonne CO_2 equivalent (CO_2eq) is the common unit used to express the warming powers of the different greenhouse gases. We sometimes find quantities of CO_2 expressed in carbon tonnes, and then we do not count the weight of the oxygen atoms. One tonne of carbon is the equivalent of 3.7 tonnes of CO_2 (the exact ratio is 44/12, since each of the two oxygen atoms has an atomic weight of 16 and the carbon atom has an atomic weight of 12).

5. At standard conditions for temperature and pressure, a tonne of pure CO_2 occupies a volume equivalent to an 8-metre-high cube, i.e., about the volume of a 200-square-metre house. Once diluted in the atmosphere at its present concentration of 380 parts per million (ppm) a tonne of CO_2 is contained in a 112-metre tall cube of air, i.e., a volume equal to that occupied by the Grande Arche de la Défense near Paris, or half the Empire State Building in New York.

The Earth's carbon

6. The Earth's carbon is divided into four main stocks: 85 per cent in the oceans (dissolved in the water or in the form of sediments), 8 per cent beneath the ground (minerals and fossil fuels) and 5 per cent in the biosphere (living organisms, organic matter in the soil). The atmosphere contains the remaining 2 per cent.

7. The amount of carbon currently stored by vegetation and the soil is 8,000 billion tonnes CO_2 equivalent, i.e., 160 times global emissions in 2007. Only a third of carbon is found directly in plants, with the remaining two-thirds stored in the soil. The quantity of carbon stored in fossil form is twice as high, amounting to 15,000 billion tonnes CO_2eq, or 300 times global emissions in 2007.

8. Peat bogs are wet soils characterized by a high proportion of organic matter of plant origin, which began forming more than 300 million years ago. They hold nearly a third of the total carbon stored in soil, even though they cover only 3 per cent of the land area. Due to human activity, these carbon stocks are declining, at a net emission rate of around 2 billion tonnes of CO_2 a year.

9. Human activity is responsible for a net flow of carbon from buried fossil fuels and the biosphere to the atmosphere and oceans. Since 1950, human activity has raised the amount of CO_2 in the atmosphere by 30 per cent and in the sea by 0.3 per cent.

10. A number of techniques are currently being studied for storing CO_2 from the combustion of fossil fuels in natural reservoirs. The most developed of these technologies involves pumping CO_2 deep inside 'empty' oil and natural gas wells. It is estimated that between 1 and 9 kilograms of CO_2 per cubic metre

can be stored in this type of reservoir. Thus around 200 cubic metres is needed to store a tonne of CO_2. Initial estimates give a global storage potential ranging from 500 to 1,500 gigatonnes of CO_2, or fifteen to fifty years' combustion of fossil fuels at present rates. Other technologies being studied include oceanic storage, mineral storage, deep aquifers and absorption onto coal, for a storage potential that could exceed 10,000 gigatonnes of CO_2.

11. It is difficult to estimate the quantity of CO_2 sequestered in a tree. It all depends on its size, the type of forest, the tree species, its age, etc. Nevertheless we can say that on average a tree sequesters 3.7 tonnes of CO_2 over 100 years, a third of which goes directly into the ground. An elm will sequester 4 tonnes in 100 years, a giant sequoia nearly 30 tonnes. This storage is not permanent. At the end of the tree's life, the carbon will be released into the atmosphere (through decomposition, burning, etc.).

Climate change and the water cycle

12. Water in its various forms is very inequitably distributed on the Earth. Nearly all of it is found in the oceans (97 per cent) and is therefore salt water. The rest is divided between the polar ice sheets and glaciers (2 per cent), underground water, lakes and soils (less than 1 per cent), and the atmosphere (less than a thousandth of 1 per cent).

13. Nearly all the world's ice lies on land, with only 2 per cent floating on the surface of the oceans (principally two ice barriers and the seasonal ice sheet in the Antarctic and all of the Arctic). If this ice melts, it makes no difference to the sea level. Only the melting of land-based ice results in raised sea levels.

14. In total, the land-based ice represents a huge ice cube 300 kilometres high. Around 97 per cent of this ice is concentrated in Antarctica and Greenland, with the rest found in glaciers and permanent snowfields. If all this were to melt, the sea level would rise by nearly 70 metres (50 centimetres for the glaciers and permanent snow, 7 metres for the Greenland ice and 60 metres for the Antarctic ice).

15. The amount of water in this ice is the equivalent to more than 4,000 years' flow of the River Amazon – the river with the largest flow rate in the world, at 220,000 cubic metres a second. At present, the Greenland ice-sheet is melting at a rate of 100–200 cubic kilometres a year, equivalent to a flow somewhere between that of the Danube and the Rhone.

16. A 1-metre rise in the sea level would result in the loss of around 2.2 million square kilometres of land, more than four times the area of France. This rise would also lead to the forced movement of 150 million people. Some parts of the

globe are more at risk than others: nearly half the land area of Bangladesh is less than 1 metre above sea level.

Impacts of climate change on the biosphere

17. Between 1945 and 2000, the harvesting dates of Châteauneuf-du-Pape wine in the Rhone valley advanced by twenty days. This same tendency has been observed in most French wine-growing areas, where on average grapes are harvested two weeks earlier than they were at the end of the second world war. All the vine growth stages occur earlier, except for the onset of dormancy, which is later due to a lack of cold days. At the same time the quantity of sugar in the grapes has increased due to higher temperatures during ripening.

18. White storks, which until the end of the 1970s migrated, are now tending to overwinter in the south and south-west of France. This tendency has increased since the end of the 1990s. The return of nesting birds also is occurring sooner – sometimes as early as January. In Spain, a large increase in the overwintering population of white storks has also been recorded.

Carbon in everyday life

19. Heating a standard house for four people over a year in France results in around 8 tonnes of CO_2 emissions if domestic heating oil is used, against 6 tonnes for gas. These emissions can be cut by 75 per cent for the same dwelling complying to high Energy Efficiency Standards.

20. Someone going away on holiday in France emits 40 grams of CO_2 per kilometre if travelling by train (ranging from 10 to 100 grams depending on the fuel used to generate the electricity), 180 grams per kilometre if travelling alone by car (100 to 400 grams depending on the model), and 270 grams per kilometre by plane (medium haul and 75 per cent occupancy rate). Thus a traveller's emissions range from a factor of one if going by train, to a factor of three if alone in a car and a factor of six if flying.

21. A small city car such as a Peugeot 206 consumes 4.5 litres per 100 kilometres and emits around 115 grams of CO_2 per kilometre, or 1.7 tonnes for 15,000 kilometres covered in the course of a year. A large sports utility vehicle (SUV) such as a Toyota Land Cruiser consumes 16 litres per 100 kilometres and emits around 400 grams of CO_2 per kilometre, or 6 tonnes for 15,000 kilometres driven over a year.

22. If one had to pay €20 per tonne of CO_2 emitted, the driver of the Peugeot 206 would face an annual cost of €34 for their emissions. The driver of the Toyota Land Cruiser would have to pay €120 a year. Given the carbon content of a litre of petrol (around 2.3 kilograms of CO_2 per litre), a carbon price of €20 per tonne of CO_2 would increase the price of petrol by 4.6 centimes a litre and of diesel by 5.2 centimes.

23. A barrel of oil contains around 420 kilograms of CO_2. If the oil producer were made to pay the price of the CO_2 content, this would represent €8.4 per barrel. It would amount to 34 per cent of the price of oil at €25 a barrel, but only 8 per cent if the oil price reached €100 a barrel.

24. Eating a meal composed of a chicken leg, 200 grams of fresh green beans, a plain yoghurt and a litre of tap water results in 600 grams CO_2 equivalent. On the other hand a meal consisting of 150 grams of beef, 200 grams of frozen green beans, a quarter of a pineapple (air-freighted from Cote d'Ivoire) and a litre of mineral water would result in emissions of around 6 kilograms CO_2 equivalent, or ten times as much.

25. A standard-capacity (800 MW) coal-burning power station operating 7,000 hours per year emits around 5 million tonnes of CO_2. If it operated only at peak periods of high demand, it would emit about 1 million tonnes of CO_2.

    A standard-capacity (400 MW) gas-fired power station operating 7,000 hours per year emits around 1.2 million tonnes of CO_2. If it operated only at periods of peak demand, it would emit about 250,000 tonnes of CO_2.

26. The average European household consumes about 5 megawatt hours (MWh) of electricity a year, which costs €600 and 'contains' around 3 tonnes of CO_2. Two-thirds of this electricity is used for heating purposes, hot water and lighting. The rest (1.7 MWh – 1 tonne of CO_2) is used for 'specific' electrical consumption, i.e. household appliances.

27. If this European household were to pay for all the carbon emitted to produce its electricity, it would pay an extra €60 a year if CO_2 were priced at €20 per tonne, or 10 per cent of its electricity bill.

28. In Europe, the price per MWh of electricity on the wholesale market is around €120 in off-peak periods. If coal is used to produce this electricity, each MWh 'contains' around 0.9 tonnes of CO_2, as against 0.5 with gas. An increase of €10 in the CO_2 price results in an increase of about 15 per cent of the wholesale price of electricity if it is completely passed on to the customer. The cost of production for the coal-using electricity producer is affected much more than for the gas-using producer.

29. A standard (blast furnace) steelworks producing a million tonnes of steel a year emits 1.8 million tonnes of CO_2, or about 2 tonnes of CO_2 per tonne of steel produced. A steelworks of the same capacity using an electric arc furnace for scrap melting emits around 0.5 tonnes of CO_2 per tonne of steel, mainly through its electricity purchases, i.e., a quarter of the amount.

30. A standard cement plant producing 1 million tonnes of cement annually emits around 700,000 tonnes of CO_2 a year, or 0.7 tonnes of CO_2 per tonne of cement produced. A standard glass factory with an annual capacity of 150,000 tonnes emits around 90,000 tonnes of CO_2 every year, or about 0.6 tonnes of CO_2 per tonne of glass produced.

Greenhouse gas emissions around the world

By Henri Casella[1]

[1] Henri Casella is a researcher at the Climate Economics Chair (CEC) at University Paris-Dauphine.

Greenhouse gas (GHG) emissions in 2005

Country	Global GHG emissions, including forestry (MtCO$_2$e)	Population (billions)	GDP per capita (current US $)	Energy consumption per capita (Toe)
China	7,218	1,304	1,710	1.4
US	6,948	296	42,708	7.8
European Union (27)	5,332	492	27,986	3.7
Brazil	2,856	186	4,787	1.2
Indonesia	2,045	219	1,300	0.8
Russia	2,021	143	5,326	4.7
India	1,877	1,095	691	0.5
Japan	1,397	128	35,633	4.1
Canada	808	32	35,205	8.3
Mexico	694	103	8,168	1.7
South Korea	609	48	17,551	4.5
France	573	61	35,105	4.4
Australia	570	21	34,716	5.9
Iran	560	69	2,746	2.4
Ukraine	495	47	1,843	2.9
Nigeria	458	141	797	0.7
Venezuela	454	27	5,423	2.3
South Africa	434	47	5,176	2.7
High-income countries	16,612	1,022	33,520	5.4
Middle-income countries	24,099	4,598	2,069	1.1
Low-income countries	1,735	823	497	0.4
World	44,130	6,444	6,954	1.8

Source: WRI, IMF, World Bank, UNFCCC.

* Land use and land use change emissions not available.

** Land use and land use change emissions from national inventory sent to UNFCCC.

GHG emissions per capita			Emissions growth between 1990 and 2005	
All gases, including forestry (tCO$_2$e / hab.)	CO$_2$ from fuel consumption (tCO$_2$ / hab.)	Agriculture, forest and waste (tCO$_2$e / hab.)	All GHGs, including forestry (%)	CO$_2$ from fuel consumption (%)
5.5	3.9	1.0	88	119
23.5	20.1	1.7	16	18
10.8	8.7	0.4**	−6	−2
15.3	1.9	13.2	13	64
9.3	1.6	7.4	14	116
14.1	10.8	1.6	−33	−32
1.7	1.1	0.5*	69	86
10.9	9.9	−0.3**	14	11
25.0	17.5	5.1	21	26
6.7	4.0	1.6	37	33
12.7	10.6	0.7*	85	102
9.4	6.8	0.7**	2	8
26.6	18.6	9.2**	39	49
8.1	6.1	0.8*	126	127
10.5	6.5	0**	−48	−56
3.3	0.7	2.1	34	45
17.1	5.7	9.4	16	30
9.2	7.3	1.4*	27	29
16.2	13.5	1.5	17	22
5.2	2.9	1.9	24	39
2.0	0.5	1.3	17	14
6.8	4.2	2.0	20	35

Sources used

The emissions data used for the table come from the Climate Analysis Indicators Tools (CAIT) program developed by the World Resources Institute (WRI). This application combines information taken from various databases, with the emphasis on sources based on scientific studies.

The CAIT database does not provide complete information about emissions related to forests or land use change. When such information was not complete we supplemented it with the information sent by countries to the United Nations Framework Convention on Climate Change (UNFCCC) secretariat, when this was available.

The breakdown among countries by income level is the one used by the World Bank in the 2010 report on world development on the basis of per capita income thresholds from $975 to $11,905 at 2008 exchange rates.

Guide to reading the table

The first column gives the total anthropogenic greenhouse gas emissions, including forests, for the seventeen largest emitters, with the twenty-seven-member European Union being considered as a single emitter, plus France.

Energy-related carbon dioxide (CO_2) emissions cover all CO_2 emitted during the production and use of energy: the production of heat and electricity, the consumption of energy derived from fossil fuels in industry, buildings and transport, fugitive emissions and all other energy uses. Emissions resulting from international air and maritime transport are included. To obtain the total energy emissions considered in Chapters 3 and 4, add in fugitive methane emissions from coal mines and oil and natural gas installations.

Greenhouse gas emissions from agriculture, forestry and waste treatment have been aggregated in the table and correspond to the agroforestry system covered in Chapters 5 and 6. This data is much less exact than the data on energy-related CO_2 emissions.

Emissions linked to **agriculture** mainly concern methane and nitrous oxide emissions arising from agricultural and stock-rearing practices. They do not include emissions from farmers' use of energy classified under energy-related emissions. The same goes for emissions resulting from waste management.

Emissions linked to **forestry and land use change** cover all net emissions resulting from the use of soil and the management of its tree cover. They can be positive or negative. The two main sources of emissions are deforestation and the draining of wetlands (especially peat bogs). Some data on land use is missing, indicated by * in the table. For Annex I countries for which data is missing in the CAIT database, we have used the figures taken from the inventories submitted to UNFCCC (indicated by ** in the table).

Emissions of energy-related CO_2 and from the agroforestry system (land-fill sites included) account for more than 90 per cent of the world's greenhouse gas emissions. The remainder come from various industrial processes (cement, aluminium, chemicals, fluorinated gases, etc.) and fugitive emissions from the energy sector (mainly methane escaping from coal mines and from oil and gas installations).

Further information

For further information on the sources used in the table and to obtain updated versions, go to the CEC website: www.prec-climat.org.

Glossary

Compiled by Anaïs Delbosc[1]

Adaptation. All actions taken to adapt society to the new environmental conditions created by climate change in order to limit the damage and maximize the benefits.

Annex I countries and Annex B countries. United Nations Framework Convention on Climate Change (UNFCCC) Annex I countries comprise developed countries and countries in transition to a market economy. They include the majority of the Kyoto Protocol's Annex B countries, that is, countries having put figures on their emissions reduction commitments. Croatia, Liechtenstein, Monaco and Slovenia come within Annex B but not Annex I. Belarus and Turkey, which are members of Annex I, are not included in Annex B.

Anthropogenic greenhouse gas emissions. Greenhouse gas emissions resulting from human activity as opposed to those occurring within the natural carbon cycle. They result mainly from the use of fossil fuels for energy production, agricultural practices, deforestation and some industrial processes.

Average cost and marginal cost. Average cost refers to the ratio of total costs to the quantity produced. Marginal cost is the additional cost resulting from the last unit produced. On account of the law of diminishing returns, economists consider that the marginal cost tends to increase when production volumes are high (and vice versa).

---

[1] Anaïs Delbosc is a project manager at CDC-Climat Recherche. She is the co-author with Christian de Perthuis of *Idées reçues sur le changement climatique* (2009).

Biomass. All organic matter of plant or animal origin. By extension, the term biomass refers to the energy that can be recovered from different organic materials (wood, waste, crops), in the form of heat, electricity and fuel.

Carbon sinks. Reservoirs that can absorb more carbon than they emit. Forests, subsoils and the oceans can all be carbon sinks. Some technologies can increase their storage capacity (e.g. carbon capture and storage on industrial sites, ocean nourishment, etc.).

Certified Emission Reduction (CER). Offset generated by one tonne of greenhouse gas reduction through a CDM-related project.

Clean Development Mechanism (CDM). Instituted by Article 12 of the Kyoto Protocol, the CDM aims to implement emissions reduction projects in developing countries (not in Annex I). The project developer obtains a CER offset for each tonne of greenhouse gas emissions avoided.

Climatology. The science of the study of successive weather conditions over long periods of time. The study of short-term weather conditions is the field of meteorology.

CO_2 equivalent (CO_2eq.). Method of measuring greenhouse gas emissions that takes into account the warming power of each gas compared to that of CO_2. This can also be expressed in terms of carbon equivalent (C): 1 kilogram CO_2eq. = 0.27 kilogram C.

Conference of the Parties (COP). Annual meeting of the signatory states of the UNFCCC, during which evolving international climate policy is negotiated.

Copenhagen Accord. A political accord that was reached at the end of the COP 15 in Copenhagen. This accord has been supported by 119 parties but has not been adopted by the COP as it only accepts unanimous decisions.

Discount rate. The rate enabling the current cost of a future benefit or future damage to be estimated. In the case of climate change, the discount rate is higher or lower depending on whether one considers that the technological possibilities for mitigation of and adaptation to climate change will increase (high discount rate) or whether one considers that the damage suffered by future generations has the same importance as for us (low discount rate).

Emissions allowance. Legally defined unit in emissions trading markets, representing one tonne of greenhouse gas. Greenhouse gas emitters subject to the market must periodically surrender to the regulatory authority as many allowances as their actual emissions in order to be in compliance.

Emission Reduction Unit (ERU). Offset generated by one tonne of greenhouse gas emissions reduction through a Joint Implementation (JI)-related project.

Energy mix. The proportion of different primary energy types used in an energy production system. An energy mix emits greenhouse gases in proportion to the fossil fuels used – the greater the proportion of fossil fuel, the greater the emissions.

European Union Emissions Trading Scheme (EU ETS). The European CO_2 emissions cap-and-trade market establishing an emissions cap for more than 11,000 industrial installations.

Forest. According to the Kyoto Protocol, a forest is considered to be any stretch of land of minimum area of 0.05 to 1 hectare, with cover at a minimum density of 10–30 per cent and minimum height of 2–5 metres. Each country is then free to specify these criteria more precisely as it sees fit.

Fossil fuels. Fuels formed by the progressive fossilization over geological time of (mainly plant) organic matter. These fuels include oil, natural gas and coal. Fossil fuels are limited in quantity and are not renewable on a human time scale.

Geo-engineering. The range of techniques that may possibly be used to directly influence the functioning of the climate system, for example by reducing the intensity of the sun's radiation reaching the Earth or by altering the functioning of natural carbon sinks.

Greenhouse gases. Gaseous constituents of the atmosphere, both natural and anthropogenic, that absorb and re-emit infrared radiation. The six anthropogenic greenhouse gases recognized by the Kyoto Protocol are carbon dioxide (CO_2), methane (CH_4), nitrous oxide (N_2O) and the fluorinated gases (SF_6, PFC, HFC). CFC fluorinated gases are covered by the Montreal Protocol.

Gross Domestic Product (GDP). Measure of the wealth created by a country. The use of purchasing power parity (PPP) enables comparisons to be made between countries, that are unaffected by abrupt changes in currency exchange rates.

Intergovernmental Panel on Climate Change (IPCC). A research group managed by the World Meteorological Organization and the United Nations Environmental Programme (UNEP), with responsibility for collecting and summarizing scientific work on climate change.

Joint Implementation (JI). Instituted by Article 6 of the Kyoto Protocol, JI promotes emissions reduction projects in developed countries (of Annex I) financed by another Annex I country. The project developer receives an ERU for each tonne of greenhouse gas emissions avoided.

Mitigation. All actions taken to reduce the concentration of greenhouse gases in the atmosphere by limiting emissions and increasing the capacity for carbon storage underground or in the biosphere and oceans.

Parts per million (ppm). Measurement unit for the concentration of greenhouse gases in the atmosphere. It states the number of greenhouse gas molecules present for every million molecules comprising the air.

Reducing Emissions from Deforestation and Forest Degradation (REDD). An internationally negotiated mechanism for financing avoided deforestation in tropical countries. REDD+ is a more comprehensive version.

Registries. National registries in greenhouse gas emissions trading schemes enable emission allowances to be monitored from their initial allocation through to their return to the regulatory authority. They also count verified emissions from regulated emitters.

Renewable energy. Energy that is permanently replenished at a rate equal to its consumption. Renewable energy sources include solar power, wind power, hydro-power, and geothermal energy. Biomass is a source of renewable energy when it comes from plants that are renewed.

Rent. Surplus income arising from the possession of a scarce good or of a particular non-reproducible asset (scarcity rent) or the occupation of an advantageous or strategic situation (economic rent). Differential (or Ricardian) rent is the surplus income arising from a structural difference in the marginal cost of production, for example resulting from differences in the fertility of agricultural land.

Tonne of oil equivalent (TOE). Measure of energy corresponding to the energy contained in a tonne of oil. One TOE is equal to the energy contained in 1.6 tonnes of high quality coal, 2.5 tonnes of lignite (poorer quality coal), 6.8 steres (cubic metres) of wood or 11.6 megawatt hours (MWh) of electricity.

United Nations Framework Convention on Climate Change (UNFCCC). The Climate Convention was signed at the Earth Summit in Rio de Janeiro in 1992 by most countries. It acknowledges the reality of climate change, aims to prevent dangerous human interference with the climate and specifies the common but differentiated responsibilities of countries.

Index